T0290855

Environmental Science, Engineering and Technology

Ecological Revitalization and Green Remediation of Contaminated Sites

Environmental Science, Engineering and Technology

Additional books in this series can be found on Nova's website under the Series tab.

Additional E-books in this series can be found on Nova's website under the E-books tab.

ENVIRONMENTAL SCIENCE, ENGINEERING AND TECHNOLOGY

ECOLOGICAL REVITALIZATION AND GREEN REMEDIATION OF CONTAMINATED SITES

ERIC S. LÓPEZ
EDITOR

Nova Science Publishers, Inc.
New York

For permission to use material from this book please contact us:
Telephone 631-231-7269; Fax 631-231-8175
Web Site: http://www.novapublishers.com

Additional color graphics may be available in the e-book version of this book.

Library of Congress Cataloging-in-Publication Data
Ecological revitalization and green remediation of contaminated sites /
editor, Eric S. López.
p. cm.
Includes index.
ISBN 978-1-61122-520-4 (hardcover)
1. Hazardous waste site remediation. 2. Bioremediation. I. López, Eric
S.
TD1052.E25 2010
628.5--dc22
2010042606

Published by Nova Science Publishers, Inc. † New York

CONTENTS

PREFACE

Ecological revitalization refers to the process of returning land from a contaminated site to one that supports a functioning and sustainable habitat. Although the final decision on how a property is reused is inherently a local decision that often rests with the property owner, the U.S. Environmental Protection Agency actively supports and encourages ecological revitalization, when appropriate, during and after the assessment and cleanup of contaminated properties under its cleanup programs. This book provides an overview of the EPA's cleanup programs and resources available to support ecological revitalization and addresses technical considerations to help cleanup project managers and other stakeholders carry out ecological revitalization at contaminated properties.

Chapter 1- Ecological revitalization refers to the process of returning land from a contaminated state to one that supports a functioning and sustainable habitat. Although the final decision on how a property is reused is inherently a local decision that often rests with the property owner, the U.S. Environmental Protection Agency (EPA) actively supports and encourages ecological revitalization, when appropriate, during and after the assessment and cleanup of contaminated properties under its cleanup programs. This document (1) provides an overview of EPA's cleanup programs and resources available to support ecological revitalization; (2) addresses technical considerations to help cleanup project managers and other stakeholders carry out ecological revitalization at contaminated properties; and (3) presents general planning and process considerations for ecological revitalization of wetlands, streams, and terrestrial ecosystems as well as successful long-term stewardship. Appendix A at the end of the document presents additional case studies on ecological revitalization.

Chapter 2- As part of its mission to protect human health and the environment, the U.S. Environmental Protection Agency (EPA or "the Agency") is dedicated to developing and promoting innovative cleanup strategies that restore contaminated sites to productive use, reduce associated costs, and promote environmental stewardship. EPA strives for cleanup programs that use natural resources and energy efficiently, reduce negative impacts on the environment, minimize or eliminate pollution at its source, and reduce waste to the greatest extent possible in accordance with the Agency's strategic plan for compliance and environmental stewardship (U.S. EPA Office of the Chief Financial Officer, 2006). The practice of "green remediation" uses these strategies to consider all environmental effects of remedy implementation for contaminated sites and incorporates options to maximize the net environmental benefit of cleanup actions.

These are an edited, reformatted and augmented versions of a United States Environmental Protection Agency publications.

In: Ecological Revitalization and Green Remediation ... ISBN: 978-1-61122-520-4
Editor: Eric S. López

Chapter 1

ECOLOGICAL REVITALIZATION: TURNING CONTAMINATED PROPERTIES INTO COMMUNITY ASSETS

United States Environmental Protection Agency

EXECUTIVE SUMMARY

Ecological revitalization refers to the process of returning land from a contaminated state to one that supports a functioning and sustainable habitat. Although the final decision on how a property is reused is inherently a local decision that often rests with the property owner, the U.S. Environmental Protection Agency (EPA) actively supports and encourages ecological revitalization, when appropriate, during and after the assessment and cleanup of contaminated properties under its cleanup programs. This document (1) provides an overview of EPA's cleanup programs and resources available to support ecological revitalization; (2) addresses technical considerations to help cleanup project managers and other stakeholders carry out ecological revitalization at contaminated properties; and (3) presents general planning and process considerations for ecological revitalization of wetlands, streams, and terrestrial ecosystems as well as successful long-term stewardship. Appendix A at the end of the document presents additional case studies on ecological revitalization.

Ecological Revitalization Under EPA Cleanup Programs Ecological revitalization of contaminated properties is consistent with EPA's mission to protect human health and the environment, and it is an integral component of EPA's cleanup programs. Under its cleanup programs, EPA ensures that (1) ecological revitalization does not compromise the protectiveness of the cleanup and (2) the best interests of stakeholders are considered. EPA's cleanup programs have established initiatives that support ecological revitalization and provide a variety of tools, information resources, and technical assistance. Collaboration and coordination with stakeholders is important for promoting ecological revitalization across EPA's programs.

Technical Considerations for Ecological Revitalization. Technical considerations for ecological revitalization include selecting appropriate cleanup technologies, addressing waste left in place, and minimizing ecological damage during the cleanup. When selecting a cleanup technology, the following may reduce ecosystem impacts during cleanup:

- Preventing access by animals that could cause damage to a cleanup technology
- Locating equipment and utilities to minimize disruption to on-site and surrounding habitat
- Selecting surface vegetation that will thrive and not interfere with the cleanup
- Evaluating the effects of amendments

Excavation and earthmoving equipment can significantly disrupt existing habitat during cleanup. Cleanup project managers are encouraged to consider the following steps to minimize habitat effects and encourage successful ecological revitalization:

- Developing and communicating ecology awareness
- Designing property-wide work zones and traffic plans
- Minimizing excavation and retaining existing vegetation
- Phasing work to stabilize one area of the property before another is disturbed
- Considering property characteristics
- Protecting on-site fauna
- Locating and managing waste and soil piles to minimize erosion
- Designing containment systems with habitat considerations
- Reusing indigenous materials whenever practical
- Controlling erosion and sedimentation
- Ensuring that borrow areas minimize effects on habitat
- Avoiding the introduction of new sources of contamination or undesirable species

For properties where waste is left in place, this document provides solutions and considerations for certain ecological revitalization issues that may arise. These include restoring soils, stabilizing metals, maintaining surface vegetation, and managing attractive nuisance issues.

Wetlands Cleanup and Restoration. Wetlands are of particular concern because in addition to intercepting storm runoff and removing pollutants, they provide food, protection from predators, and other vital habitat factors for many of the nation's fish and wildlife species. Important considerations for planning and designing wetland cleanup and restoration include:

- Evaluating the characteristics, ecological functions, and condition of wetlands
- Determining beneficial wetland functions and structures after the cleanup
- Developing a wetlands design that will achieve the stated ecological functions
- Ensuring that cleanup activities and wetland features have minimal effects on existing wetlands
- Specifying and implementing explicit maintenance requirements

Stream Cleanup and Restoration. Stream cleanups often disrupt stream flow and habitat. Considerations for (1) designing and implementing cleanups that facilitate ecological revitalization of streams and stream corridors and (2) mitigating adverse ecological effects of constructing cleanup features include:

- Stream channel restoration decisions about channel width, depth, cross-section, slope, and alignment
- Streambank stabilization measures (temporary and permanent)
- Streambank vegetation approaches
- Management of watershed processes such as increased runoff or sediment loading from construction

Bioengineering techniques that stabilize the soil or streambank by establishing sustainable plant communities have become an increasingly popular approach to streambank restoration. Stabilization techniques may include using a combination of live or dormant plant materials, sometimes in conjunction with other materials such as rocks, logs, brush, geotextiles, or natural fabrics.

Terrestrial Ecosystems Cleanup and Revitalization. Establishing a plant community that will thrive with minimal maintenance is a critical step in developing a healthy terrestrial ecosystem on cleanup properties. Factors to consider when establishing terrestrial plant communities in disturbed areas include:

- Soil suitability and the need for soil amendments or soil stabilization
- Property-specific plant selection with a preference for native plants
- Protection from disturbances (such as from grazing animals and vehicles)
- Timing to ensure optimal plant establishment

Long-Term Stewardship Considerations. On cleanup completion, operation and maintenance (O&M) activities through responsible stewardship protect the integrity of the cleanup and the functioning of the associated ecosystems. Specifically for properties where waste is left in place, long-term stewardship is necessary to ensure protectiveness of the remedy. When designing a successful O&M program for ecological revitalization, it is important to consider the following:

- Planning early for long-term stewardship
- Incorporating ecological revitalization components into general maintenance activities
- Establishing a monitoring program that incorporates the ecological revitalization components
- Using institutional controls to prevent activities that could potentially interfere or disturb ecologically revitalized areas

1.0. Introduction

Revitalizing properties for ecological purposes helps to achieve U.S. Environmental Protection Agency (EPA)'s goal of restoring contaminated properties to environmental and economic vitality. The term "ecological revitalization" refers to the process of returning land from a contaminated state to one that supports functioning and sustainable habitat. Although the final decision on how stakeholders will reuse a property is inherently a local decision that often rests with the property owner, EPA supports and encourages ecological revitalization as part of the cleanup of contaminated properties across all of its cleanup programs. Ecological revitalization has many positive effects that apply to a variety of stakeholders (see text box below). The objectives of ecological revitalization and those of the remediation process are best accomplished if they are carefully coordinated. To this end, this document provides general information for coordinating ecological revitalization during the cleanup of contaminated properties, as well as technical considerations for implementing ecological revitalization of wetlands, streams, and terrestrial ecosystems during cleanup.

The purpose of this document is to assist cleanup project managers and other stakeholders to better understand, coordinate, and carry out ecological land revitalization at contaminated properties during cleanup. The focus of this document is primarily on planning-level issues, not detailed design approaches, along with technical information and references for executing ecological revitalization activities at contaminated properties. This document highlights (1) several considerations and initiatives under EPA's Office of Solid Waste and Emergency Response (OSWER) cleanup programs that support ecological revitalization, (2) a variety of tools and resources that are available to assist cleanup project managers and other stakeholders, and (3) case studies that provide examples of ecological revitalization at cleanup properties. Another purpose of this document is to help facilitate cross-program networking while planning, designing, and implementing cleanups to help increase valuable ecosystems that are created or improved through ecological revitalization. To that end, Appendix A provides case studies on ecological revitalization approaches taken at various cleanup properties and identifies specific points-ofcontact who can provide valuable insights for those interested in implementing ecological revitalization at their properties.

Ecological Revitalization Benefits a Variety of Stakeholders

Cleanup Project Managers. A restored habitat can reduce long-term operation and maintenance (O&M) requirements without compromising the effectiveness of the cleanup action. A restored habitat can also help optimize property engineering controls, such as using vegetation to reduce surface water infiltration or using wetlands as part of stormwater controls.

Potentially Responsible Parties. A valuable restored habitat could enhance a company's image and reputation in the community. Getting a property cleaned up and reused can also ease liability concerns, which in turn may have a positive financial impact.

Local Government. An ecological reuse may increase tourism, tax revenues, property values, and quality of life for residents.

Local Citizen Groups and Individuals. Increasing habitat and passive recreational activities can improve the character of the neighborhood, employment opportunities, and area air and water quality.

Environmental Organizations. Ecological revitalization projects may provide the opportunity to protect or improve local and regional habitats.

The document is organized into the following sections:

- **Section 2** presents an overview of EPA's cleanup programs and their revitalization initiatives, tools, and resources available to support ecological revitalization.
- **Section 3** provides general technical considerations for implementing ecological revitalization, including cleanup technology considerations, cleanup planning and design issues, and considerations for minimizing ecological damage during cleanups.
- **Section 4** provides technical considerations for planning and designing wetland cleanups and restoration efforts.
- **Section 5** provides technical considerations for designing and implementing cleanups that facilitate ecological reuse of streams and stream corridors and for mitigating potential adverse ecological impacts of constructing cleanup features.
- **Section 6** presents factors to consider for establishing terrestrial plant communities in disturbed areas, including general revegetation principles; protecting or creating natural terrestrial ecosystems, meadows, or prairies; and establishing vegetation on semi-arid or arid lands.
- **Section 7** provides considerations for operation and maintenance (O&M) activities to ensure the ongoing integrity of the cleanup and functioning of the associated ecosystems after cleanup completion.

This document was developed by EPA's OSWER cleanup programs, including the Office of Superfund Remediation and Technology Innovation (OSRTI), Office of Resource Conservation and Recovery (ORCR) (formerly known as Office of Solid Waste), Federal Facilities Restoration and Reuse Office (FFRRO), Office of Brownfields and Land Revitalization (OBLR), and Office of Underground Storage Tanks (OUST) (see the OSWER organizational chart, shown on page iii). **Figure 1-1** on the following page identifies specific elements of each OSWER program office's strategic plans, action plans, or program policies that establish support for ecological revitalization. EPA also encourages other public and private interests, including state and local governments and land trusts, land banks, and nonprofit organizations to participate in ecological revitalization activities, particularly in long-term stewardship at cleanup properties. While the scope of this document includes the EPA offices listed above, the information could be useful to a wide variety of additional stakeholders with an interest in the reuse or redevelopment of a cleanup property, specifically to create, restore, improve, or protect ecological resources. Therefore, this document also provides information that can be applicable to cleanup project managers, potentially responsible parties, Resource Conservation and Recovery Act (RCRA) corrective action facility owners/operators, local governments, citizen groups, environmental organizations, and other interested individuals.

1.1. Ecological Revitalization and Ecological Reuse

The terms "ecological revitalization" and "ecological reuse" are often used interchangeably. However, there is a subtle distinction between the terms. Ecological revitalization refers to *the technical process* of returning land from a contaminated state to one that supports functioning and sustainable habitat. Ecological reuse refers to the *outcome* of a cleanup process and includes those areas where proactive measures (such as a conservation easement) have been implemented to create, restore, protect, or enhance a habitat for terrestrial or aquatic plants and animals (EPA 2006e). In this sense, the process of ecological revitalization of a property can lead to an ecological reuse outcome.

Ecological reuse is different from greenspace use in that, in addition to habitat, the latter can include parks, playgrounds, and gardens; ecological reuse strives to restore native habitat and does not include active recreation activities. However, low-impact or passive recreation, such as hiking or bird watching, may occur at ecological reuse properties. In addition, ecological revitalization can occur on a portion of a cleanup property adjacent to greenspace use (for example, a golf course with native plant species surrounding the course), commercial operations, or industrial use. Further, ecological revitalization can occur at varying degrees; some areas of a property may be restored to relatively pristine, historic conditions, while other areas may be planted with native or other compatible species. Both degrees of ecological revitalization lead to habitat that one may accurately characterize as ecological reuse.

Ecological Revitalization and Ecological Reuse

There is a distinction between the terms ecological "revitalization" and "reuse" but they are related. Ecological revitalization returns land to a functioning and sustainable habitat. Ecological revitalization of a site can lead to an ecological reuse, where proactive measures have been implemented to create, restore, protect, or enhance a habitat for terrestrial or aquatic plants and animals (EPA 2006e).

1.2. General Program Initiatives

EPA's 2006-2011 Strategic Plan (EPA 2006a) restates EPA's commitment to protect human health and the environment, including restoring the nation's contaminated land and enabling communities to return restored properties safely to beneficial economic, ecological, and social use. As part of the strategic plan, EPA established five goals, including:

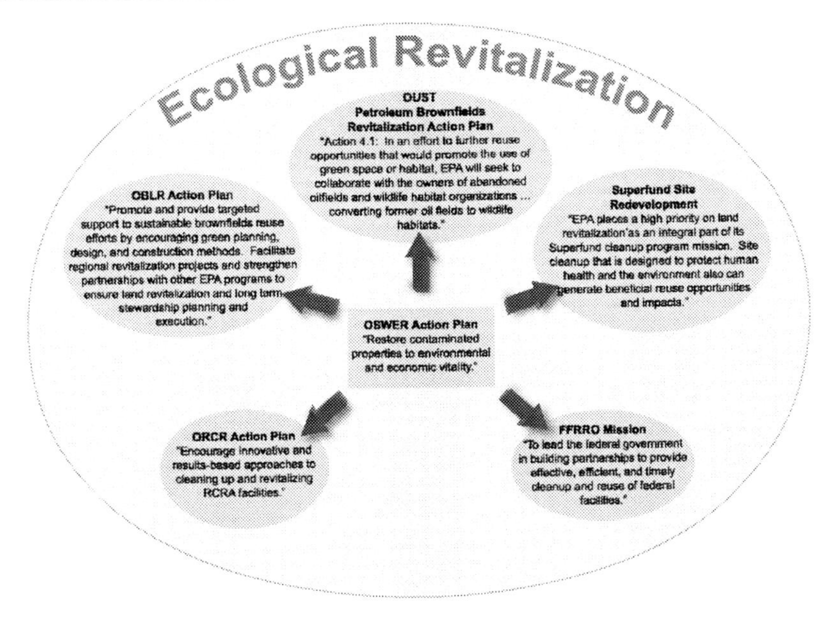

Figure 1-1. Ecological Revitalization as a Component of EPA Cleanup Programs

INTERSTATE TECHNOLOGY AND REGULATORY COUNCIL (ITRC) COLLABORATION ON ECOLOGICAL REVITALIZATION

ITRC, a state-led coalition working with the federal government, industry, and other stakeholders to achieve regulatory acceptance of environmental technologies, has compiled a wealth of information on ecological revitalization. ITRC's document "Planning and Promoting Ecological Land Reuse of Remediated Sites" (ITRC 2006) provides recommendations that are applicable to active and inactive properties and all programs. Visit the following Web site for more information: www.itrcweb.org.

- Clean Air and Global Climate Change (Goal 1)
- Clean and Safe Water (Goal 2)
- Land Preservation and Restoration (Goal 3)
- Healthy Communities and Ecosystems (Goal 4)
- Compliance and Environmental Stewardship (Goal 5)

Ecological revitalization contributes to each of these goals. For example, EPA's cleanup programs (under Goal 3) have set a national goal of returning formerly contaminated properties to long-term, sustainable, and productive use (EPA 2006a). These programs include Superfund (under authority of the Comprehensive Environmental Response, Compensation, and Liability Act [CERCLA] of 1980, as amended), Corrective Action (under authority of RCRA), Underground Storage Tanks (UST), Federal Facilities Restoration and Reuse, and Brownfields (under Goal 4). In 2003, EPA introduced the Land Revitalization Initiative to (1) promote cross-program coordination on land reuse and revitalization projects

Table 1-1. Cross-Program Revitalization Measures Tracked by Each EPA Cleanup Program

Performance Measures and Indicators	EPA Cleanup Program				
	OSRTI	ORCR	FFRRO	OBLR	OUST
Universe Indicator: The number of contaminated, potentially contaminated, or previously contaminated properties and surface acres for which OSWER's cleanup programs have an oversight role for assessment or response action.	a	b	a	c	d
Protective for People (PFP) measure: The number of acres at which there is no complete pathway for human exposures to unacceptable levels of contamination based on current property conditions.	a	b	a	c	d
Ready for Anticipated Use (RAU) measure: The number of acres at a property that meets the criteria for the PFP measure, as well as (1) all cleanup goals have been achieved for current and reasonably expected land uses and (2) all institutional or other controls have been put in place.	a	b	a	c	d
Status of Use Indicator: How the acres at a property subject to the Universe Indicator are being used at the point in time when the determination is made.	a	**	a	--	--
Type of Use Indicator: For programs, regions, states, local governments, or tribes that are looking for measures they could use to help describe in more detail how contaminated or potentially	a	**	a	c	--

contaminated properties under their jurisdiction are currently being used. For example, "ecological use" is a type of use under this indicator.					

References: EPA 2007e; f; g and EPA 2009

Notes:

** Reporting of Indicator is voluntary at this time.

-- Indicator not tracked.

a New Land Reuse Module in Comprehensive Environmental Response, Compensation and Liability Information System (CERCLIS) used to track CPRM information, independent of Government Performance and Results Act (GPRA) goals. OSRTI reports "Ready for Reuse" as a GPRA measure (based on status of cleanup and institutional controls [IC]), which equates to both PFP and RAU.

b Through 2008, the RCRA facility Indicator Universe will consist of all RCRA Corrective Action 2008 GPRA baseline facilities. For 2009 and beyond, the RCRA facility Indicator Universe will consist of all RCRA Corrective Action 2020 facilities. The Current Human Exposures Under Control Environmental Indicator (HE EI) will be used to report the PFP measure. A "RCRA RAU Documentation" form has been developed to assist in implementing this performance measure. Status of Use and Type of Use indicators are not being required at a national level. Universe and RAU data elements have been incorporated into the RCRA Information System (RCRAInfo Version 4.0 released in December 2008).

c OBLR is using Property Profile Form data to report on the Universe Indicator (properties and acres where assessment or cleanup are reported as complete for the first time under a Brownfields grant) and Type of Use Indicator (Greenspace, Residential, Commercial, Industrial, and Mixed Use). OBLR is also using their Property Profile Form to collect information on the "Ready for Reuse" measure (based on status of cleanup and IC), which equates to both PFP and RAU measures and is being reported as a Government Performance and Results Act measure by OBLR. Indicator and measure information is being tracked in the EPA OBLR Assessment, Cleanup, and Redevelopment Exchange System (ACRES) database.

d OUST's "Confirmed Release" will equal one site and one acre for the Universe Indicator; OUST's "Cleanup Completed" will equal one acre for both the PFP and RAU performance measures.

and (2) ensure that stakeholders clean up contaminated properties and make them available for productive use. At properties that involve multiple cleanup programs, land revitalization encourages a "one cleanup program" approach to improve consistency, management, and cost-effectiveness of the program. Cleaning up previously contaminated properties for reuse reinvigorates communities, preserves open space, and prevents sprawl. This initiative goes beyond ecological revitalization, and stakeholders can use land in many ways, including new public parks, restored wetlands, and new businesses. For more information on land revitalization, visit the following Web site: www.epa.gov/oswer/landrevitalization/basic information.htm.

In 2006, OSWER issued the Interim Guidance for OSWER Cross-Program Revitalization Measures (CPRM) (EPA 2006b, 2006e) to help track land revitalization at the national level. These revitalization measures show how EPA cleanup programs currently track their revitalization activities, as shown in Table 1-1.

While all environmental restoration activities that lead to reuse options are beneficial, this document focuses on ecological revitalization, which is becoming even more important as communities are increasingly seeing ecological revitalization as a desirable process to achieve a viable reuse outcome.

1.3. General Process Considerations

Ecological revitalization activities can occur on a wide variety of properties and could be compatible with several types of end uses. When considering ecological revitalization at a property, it may be useful to consider the following:

- It is important to begin the ecological revitalization process early in the cleanup.
- Ecological revitalization is not a short cut for cleanup and can have strict cleanup standards.
- Habitat can be created on an entire property or on a portion of a property, and can be created adjacent to other end uses such as intermodal centers or industrial areas.
- Ecological revitalization is not typically considered an "enhancement," so it can generally be funded by EPA (under the Superfund Program, for example), and may be needed under Section 404 of the Clean Water Act.
- Ecological revitalization provides a variety of environmental, economic, and social benefits.

The remainder of this document further discusses these considerations.

Ideally, the process of ecological revitalization begins during the assessment or investigation phase of a cleanup rather than after the remedy is underway; this allows for the greatest range of potential options and end uses. As discussed throughout this document, ecological revitalization needs additional considerations to ensure protection of wildlife that could end up inhabiting the cleaned up property, in addition to protecting human health and the environment. Some of these additional considerations are included in **Figure 1-3.**

Ecological revitalization is not a short cut for property cleanup, but rather a viable and productive reuse option that also ensures protection of human health and the environment. Potential challenges to consider early in the process include (1) liability if additional cleanup or maintenance is needed, especially in the long term; (2) public health and access if the cleanup property is converted to habitat; (3) how ecological revitalization, which can be slower than other reuse alternatives, will impact surrounding areas, and (4) transfer of land and long-term stewardship. Therefore, while ecological revitalization can be considered at all contaminated properties, it may not be appropriate for all properties. There are a variety of considerations needed to ensure protectiveness (further discussed in Section 2), including conducting an ecological risk assessment (ERA), avoiding attractive nuisances (see definition on page 3-2), and bioaccumulation issues. For example, at the Bunker Hill Superfund Site in Idaho (shown in Figure 1-2), attractive nuisance issues were taken into account while ecological revitalization was being considered as an option. For additional information on bioaccumulation and EPA's persistent, bioaccumulative, and toxic chemical program, visit the following Web site: www.epa.gov/pbt/index.htm. In addition, ecological revitalization may require other considerations to ensure successful creation of habitat, such as controlling invasive plant species. Technical performance measures (TPM) are available to determine the success of ecological revitalization as part of a cleanup process. For additional information on TPMs, visit the following Web site: www.clu-in.org/products/tpm.

Although commercial, industrial, residential, and some recreational uses are not ecological reuse, habitat can be incorporated as a portion of or adjacent to these redeveloped areas. For example, at the Joliet Army Ammunition Plant (JOAAP), a tallgrass prairie was created among large intermodal centers and other industrial areas. British Petroleum (BP) also plants native vegetation at its refineries adjacent to areas where occasional spills may occur to provide phytoremediation, if necessary. See Appendix A for additional information regarding the JOAAP in Illinois and the BP Former Refinery in Wyoming (a photograph of JOAAP revitalization is also included on the cover of this document).

Figure 1-2. Before and after photographs of the Bunker Hill Superfund Site in Idaho where contamination was left on-site and capped with biosolids compost and wood ash. A long-term O&M plan was established to ensure that attractive nuisance (see definition on page 3-2) issues did not result. See Appendix A for additional information. *Photographs courtesy of Dr. Sally Brown, University of Washington.*

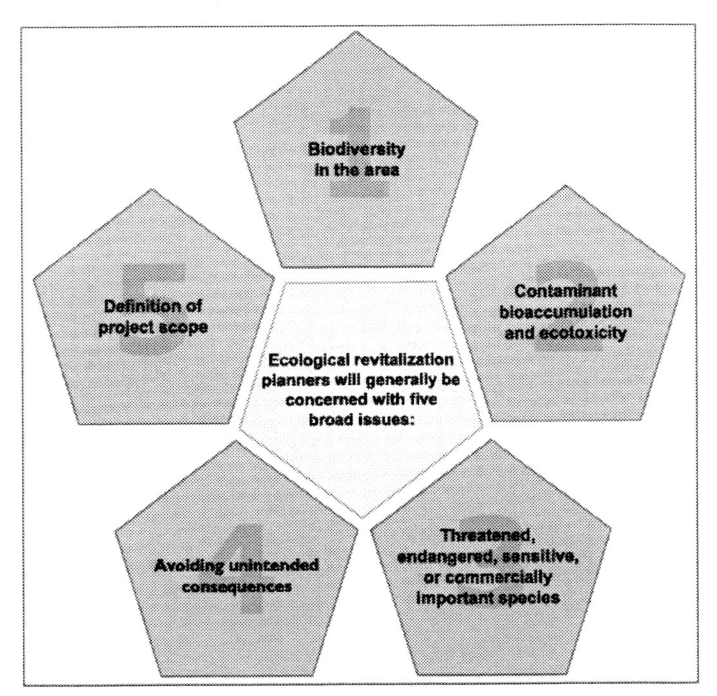

Figure 1-3. Considerations When Planning for Ecological Revitalization

Ecological revitalization provides a variety of positive environmental, economic, and social impacts. Some positive impacts of ecological revitalization are as follows (Interstate Technology and Regulatory Council [ITRC] 2006; EPA 2006d):

- Repairs damaged land
- Improves soil health
- Supports diverse vegetation
- Reduces erosion
- Sequesters carbon
- Controls landfill leachate
- Protects surface and ground water from potential contamination
- Helps remove stigma associated with prior waste site
- Enhances property values and raises tax revenue (www.epa.gov/superfund/programs/recycle/pdf/method.pdf)
- Provides passive recreational opportunities
- Contributes to a green corridor or infrastructure

Additional environmental, economic, and social impacts are listed in the ITRC's document, "Making the Case for Ecological Enhancements" at www.itrcweb.org/Documents/ECO-1.pdf.

The remainder of this document provides background information on ecological revitalization in relation to EPA's cleanup programs, and technical information and resources to assist in implementing ecological revitalization at contaminated properties.

2.0. Ecological Revitalization under EPA Cleanup Programs

EPA's mission across its cleanup programs is to protect human health and the environment. Ecological revitalization of contaminated properties is consistent with this mission and is an integral component of EPA's cleanup programs. EPA recognizes the important role that it plays in helping communities and other stakeholders clean up and reclaim contaminated properties, which has led to specific programs and initiatives that support the revitalization and reuse (or continued productive use) of properties as part of their assessment and cleanup. The nature and extent of EPA involvement in supporting ecological revitalization varies from program to program, as well as from property to property. Moreover, the decision on whether and how stakeholders will reuse a property for ecological or other purposes is inherently a local decision that usually rests with the property owner.

This section presents an overview of each cleanup program under EPA OSWER (see the organizational chart on page iii of this document) and its revitalization initiatives, which provides the programmatic context for evaluating and taking steps to support ecological revitalization as part of cleaning up contaminated properties. Section 2.1 provides several considerations that are common to each cleanup program; Sections 2.2 through 2.6 address each program separately.

2.1. General Programmatic Considerations

Depending on the specific circumstances at a contaminated property, EPA's OSWER cleanup programs manage, oversee, or provide assistance with investigation and cleanup under one of several different programs, including the Superfund, Federal Facilities, RCRA Corrective Action, Brownfields, and UST programs. In some cases, individual contaminated properties can be subject to multiple OSWER programs. For example, the Rocky Mountain Arsenal involves the RCRA Corrective Action, Superfund, and Federal Facilities programs (Appendix A provides a case study on this site; a photograph is also included on the cover of this document). As illustrated in **Table 2-1** below, a variety of property types can fall under the purview of one or more programs. With proper planning, these programs can support ecological revitalization as part of, or following, cleanup.

Ecological Revitalization Cleanup Standards in the Calumet Region, Chicago, Illinois

On the south side of Chicago, Illinois, a roundtable team of federal, state, and local agencies developed the Calumet Area Ecotoxicology Protocol to specifically address ecological revitalization activities in this region (Calumet Ecotoxicology Technical Roundtable Team 2007). The protocol includes cleanup standards that are protective for both human health and ecological receptors, which may be more stringent than federal and state industrial and commercial cleanup goals. Sites being cleaned up in the Calumet Region follow the protocol to ensure protectiveness of human health and the environment as well as streamline the cleanup process.

Whether being addressed under one or several of EPA's cleanup programs, several factors determine whether and how ecological revitalization can be supported at a specific property. These factors are discussed below.

Protectiveness. An important consideration when evaluating the ecological revitalization of a property is ensuring protectiveness for both human health and the environment. EPA does not lower its standards of protection for a property that will be reused, nor does it allow reuse to reduce effectiveness of cleanup measures. Under its cleanup programs, EPA ensures that contamination is either completely removed, cleaned up to acceptable levels, or managed using protective measures that reduce the possibility of exposure to the contamination. If all contamination is eliminated, then human health and the environment are fully protected and the land or water body is available for ecological or others types of use. Where protective measures are in place for waste that remains after the cleanup, EPA determines whether such measures will continue to provide protection for ecological reuse, or whether that use might impair the protective measures. In some cases, the presence of certain contaminants (for example, persistent pollutants that are readily bioavailable, such as metals and polycyclic aromatic hydrocarbons [PAH]) remaining after the cleanup may preclude ecological revitalization efforts on those portions. Cleanup project managers will make these determinations on a case-by-case basis. One of the key challenges to implementing ecological revitalization under EPA's cleanup programs is that cleanup goals applicable to habitat creation can necessitate complex analyses. Cleanup goals for ecological protection may also need to be more stringent than for protection of human health (see text box above). Another challenge stems from a lack of familiarity with ecological end uses and ways in which to quantify the value of such end uses (EPA 2005).

Enhancement. The extent of EPA's involvement in supporting ecological revitalization at a contaminated property depends on the cleanup program involved, the legal authorities under which the property operates, and the specific property at issue. For example, under the Superfund Program, EPA cannot fund ecological enhancements (that is, activities not necessary for the protection of human health and the environment); rather, it can encourage enhancement activities funded by other stakeholders and can fund aspects of a cleanup project that are necessary for the anticipated future uses of a property. Under the Superfund Program, EPA can fund activities to better understand the reasonably anticipated future land use, which informs remedy selection and implementation and helps support long-term protectiveness. Anticipating the future use of a Superfund site after cleanup completion is of key importance in selecting and designing a remedy that will be consistent with that use. Similarly, EPA's Brownfields Program provides, among other things, technical assistance to communities to support plans for ecological and other "green" enhancements to the cleanup and reuse of properties (for example, designing rain gardens, native landscaping, or green infrastructure), but not the actual revitalization or reuse activities themselves. Other programs, such as RCRA Corrective Action or UST, encourage and support ecological revitalization through their established relationships with states that have delegated programs and through collaborative efforts with governmental and non-governmental organizations. State programs may also have limitations for funding activities that are not directly needed for the protection of human health and the environment. Property owners may see the benefits of supporting the reuse of properties, including the ecological revitalization of the land, particularly when it affects public perception of their business operations and commitment to the environment. Moreover, EPA may be able to offer certain incentives to support ecological revitalization under its initiatives, such as EPA's Environmentally Responsible Redevelopment and Reuse (ER3) Initiative.

Empire Canyon, Daly West Mine Site, Summit County, Utah

A resort development company has proposed the construction of a hotel, spa, and condominium project at the Daly West Mine Site, to be known as the Montage Resort & Spa. The development will contribute to the cleanup of contamination at this former mining site in Park City, Utah. The developer agreed to participate in EPA's Environmentally Responsible Redevelopment and Reuse (ER3) Initiative for contaminated properties. As an ER3 participant, the Montage Resort & Spa will incorporate extensive "green" features into the design, construction, and operation of the development, including several ecological revitalization components. For example, the project involves treatment of ground water collected by foundation drains using a constructed wetland; a native vegetation management plan to improve ecosystem health and reduce the risk of wildfires around the site; and a conservation easement for 2,800 acres of open space to offset additional density from the project. By incorporating sustainable practices and principles into the project, the developer has minimized the impact of the project on the environment without sacrificing profitability.

Table 2-1. Property Types Commonly Managed under EPA Cleanup Programs

Example Property Type	EPA Cleanup Programs				
	Superfund	Federal Facilities	RCRA Corrective Action	Brownfields	UST
Foundry	X		X	X	
Gas Station				X	X
Landfill	X	X	X	X	
Manufacturing Facility	X		X	X	X
Industry/Solvent Use	X		X	X	X
Military Installation	X	X	X		X
Other Federal Facilities*	X	X	X		X
Mining	X	X		X	
Refinery	X		X	X	X
Tannery	X		X	X	

* Non-military use facilities owned or operated by the federal government

In general, most ecological revitalization efforts are not considered enhancements if the activities are necessary for the anticipated future ecological use of the property or to restore ecological function and, therefore, can be considered and incorporated into property cleanup plans. Even costs for extensive revitalization efforts to create or restore the function of an ecosystem can be justified if the revitalization is needed because of environmental stressors or adverse impacts to the property caused by the cleanup. For example, grasses, shrubs, and other native plants serve a practical function of stabilizing soil to prevent erosion, while also improving the property's aesthetics and ecological function.

Stakeholder Involvement. Regardless of which EPA program is involved in the assessment, cleanup, and revitalization of a contaminated property, numerous stakeholders may have an interest in the actions taken at the property, including the following:

- Other federal, state, local, or tribal agencies
- Parties responsible for the contamination
- Current landowners
- Neighboring property owners and the surrounding community
- Prospective purchasers or future users of the property

With different stakeholders potentially involved at a contaminated property, the ecological revitalization of the property will need to consider the varied interests, objectives, and requirements of those stakeholders. Successful ecological revitalization efforts have typically resulted from well-facilitated processes that encourage open communication and the exchange of information among the stakeholders at a property.

Additional Initiatives That Support Sustainable Cleanup and Reuse In addition to specific initiatives that are supported by EPA's cleanup programs (and described in the following sections), there are other EPA initiatives that can also support ecological revitalization at contaminated properties regardless of which OSWER program is supporting the cleanup. These initiatives include the following:

EPA's EcoTools Initiative provides a variety of resources for cleanup project managers, especially under the Superfund program. In addition to technical information, the EcoTools Web site provides cleanup project managers access to ecological experts via a technical assistance service. For more information, visit www.clu-in.org/ecotools.

EPA's ER3 Initiative uses enforcement and other EPA-wide incentives to promote sustainable cleanup and redevelopment of contaminated properties. Under the ER3, EPA collaborates with federal, state, public, and private partners to identify, develop, and deliver incentives to encourage developers and property owners to implement sustainable practices during the redevelopment of contaminated properties. The primary components of ER3 are to (1) identify and provide enforcement and EPA-wide incentives to developers and property owners to encourage sustainable cleanup and development; (2) develop partnerships with federal, state, public, and private entities to establish a network of expertise on sustainable development issues; and (3) promote sustainable redevelopment of contaminated properties through education and outreach. For more information on ER3, visit www.epa.gov/ compliance.

Other Cross-Cutting Ecological Revitalization Considerations for EPA Cleanup Programs

Liability: Consider who will be responsible if additional cleanup or maintenance is required, especially in the long-term.

Public Health and Access: Consider whether the public will safely be allowed to use the property if it is converted to habitat.

Surrounding Areas and Time: Ecological revitalization can impact surrounding areas because, while ecological revitalization can be a more cost- effective process, the time required to return a property to functioning and stable habitat can take longer than other reuse alternatives.

Transfer of Land and Long-Term Stewardship: Ensure that institutional controls are in place and operating effectively, and consider who will be the long-term landowner responsible for stewardship of the ecological revitalization and associated natural resources.

EPA's Five Star Restoration Program brings together students, conservation corps, other youth groups, citizen groups, corporations, landowners, and government agencies to provide environmental education and training through projects that restore wetlands and streams. The program provides challenge grants, technical support, and opportunities for information exchange to enable community-based restoration projects. Visit www.epa.gov/owow/wetlands/restore/5star for additional information about the Five Star Restoration Program.

EPA's GreenAcres Initiative promotes natural and sustainable landscaping practices using native plants and other green landscaping strategies. The GreenAcres Initiative is a component of EPA's Great Lakes National Program Office and its efforts to promote an integrated, ecosystem approach to protect, maintain, and restore the chemical, biological, and physical integrity of the Great Lakes. Under GreenAcres, EPA provides information and resources on using native plants and natural landscape approaches in urban, suburban, and corporate settings. For more information, visit www.epa.gov/greenacres.

EPA's Green Infrastructure Partnership is an initiative to work with partners to promote green infrastructure as an environmentally preferable approach to stormwater management. In January 2008, EPA and its partners released an action strategy for managing wet weather with green infrastructure. The strategy provides a collaborative set of actions that promote the use of green infrastructure and outlines efforts to bring green infrastructure technologies and approaches into mainstream wet weather management. For more information about this partnership and the action strategy, visit *http://cfpub.epa.gov/npdes/home.cfm?program_id=298.*

EPA's Green Remediation Initiative promotes the use of best management practices (BMP) to maximize the net environmental benefits of cleanup actions. With the help of public and private partners, EPA OSWER is documenting the state of BMPs, identifying ways to improve BMPs, and forming a community of BMP practitioners. Technical assistance is offered to cleanup project managers to find new opportunities for reducing the environmental footprint of cleanup actions. For more information about this initiative, visit www.clu-in.org/greenremediation.

EPA's GreenScapes Program identifies cost-efficient and environmentally friendly solutions for landscaping. Designed to help preserve natural resources and prevent waste and pollution, GreenScapes encourages companies, government agencies, other entities, and homeowners to make more holistic decisions regarding waste generation and disposal and the associated impacts on land, water, air, and energy use. Visit www.epa.gov/greenscapes for additional information on the GreenScapes Program.

2.2. Superfund Sites

EPA's OSRTI carries out the Superfund Program, which addresses contamination from uncontrolled releases at hazardous waste sites that threaten human health and the environment. EPA manages the Superfund Program under the authority of the CERCLA, 1980, as amended. Under the Superfund Program, abandoned, accidentally released, or illegally dumped hazardous wastes that pose a current or future threat to human health or the environment are cleaned up. To accomplish its mission, EPA works closely with communities, potentially responsible parties, and other federal, state, local, and tribal agencies. Together with these groups, EPA identifies hazardous waste sites, investigates the conditions of the sites, formulates cleanup plans, and cleans up sites to ensure that they are protective of human health and the environment.

Superfund cleanups include both long-term and short-term response actions. Long-term cleanups or remedial actions are conducted on sites that, following an evaluation, are listed on the National Priorities List (NPL). Once on the NPL, EPA follows a thorough process to carefully investigate the site and select and carry out a remedy specific to that site. Short-term cleanups called removal actions, fall into three categories: (1) non-time critical responses at sites where on-site activities do not need to be initiated for more than six months; (2) time critical responses at sites where on-site activities must begin within six months; and (3) emergency removal actions at sites that need initiation of on-site activities within hours of the decision that action is necessary. EPA's role and ability to support ecological revitalization may vary across these different site types, as discussed below.

Coordinating Ecological Revitalization Efforts in the Superfund Remediation Process. OSRTI established the Superfund Redevelopment Initiative (SRI) to ensure that at every Superfund site, EPA and its partners have the necessary tools and information to return the country's most hazardous sites to productive use, including information related to natural resources and ecological revitalization. In addition to cleaning up Superfund sites and making them protective of human health and the environment, communities and other partners are involved in considering future use opportunities and integrating appropriate reuse options into the cleanup process. At previously cleaned sites, communities are also involved to ensure the long-term stewardship of the site remedies. For more information on the SRI, visit the following Web site: www.epa.gov/superfund/programs/recycle.

When investigating, designing, and implementing a cleanup, remedial project managers (RPMs) are encouraged to consider, to the extent practical, anticipated future land uses. With careful planning, many Superfund sites can accommodate ecological revitalization while still meeting the requirements under CERCLA and other federal and state regulations. Stakeholders best accomplish the objectives of ecological revitalization and those of the remediation process through careful coordination. For example, under CERCLA EPA needs to coordinate with all affected Natural Resource Trustees (Trustees) when conducting a remedial investigation (RI). Trustees are designated under Executive Order 12580 and defined under CERCLA as other federal, state, or tribal governments that act on behalf of the public for natural resources under their trusteeship. Trustees often have information and technical expertise about the biological effects of hazardous substances, as well as the location of sensitive species and habitats that can assist EPA in evaluating and characterizing the nature and extent of site-related contamination. Coordination at the investigation and planning stages provides the Trustees early access to information they need to assess injury to natural resources. This assists Trustees in making early decisions about whether sites need restoration in light of the response actions.

Several types of ecological studies, including ERAs and Natural Resource Damage Assessments

(NRDAs), support cleanup and ecological revitalization decisions at a Superfund site. EPA utilizes an ERA as part of its process for assessing the risks of site-related contamination. ERAs are usually conducted during the Remedial Investigation/Feasibility Study (RI/FS) phase of the Superfund response process and inform RPMs about the risk associated with the site. While physical impacts of site cleanup activities are assessed during the FS, ERAs specifically evaluate the likelihood that adverse ecological effects are occurring or may occur because of exposure to chemical (for example, release of hazardous substances) stressors at a site. These assessments often contain detailed information regarding the interaction of these "stressors" with the biological community at the site. Part of the assessment process includes creating exposure profiles that describe the sources and distribution of harmful entities, identify sensitive organisms or populations, characterize potential exposure pathways, and estimate the intensity and extent of exposures at a site. The National Oceanic and Atmospheric Administration (NOAA), a natural resource trustee, and the U.S. Army Corps of Engineers (USACE) played an important role in remediation of the Atlas Tack Superfund Site in Massachusetts, including conducting a site-specific ERA (EPA 2008h) based on the cleanup goals that were established for this site (see text box on this page and **Figure 2-1**). Additional information about this remedy is available at http://www.clu-in.org/download/newsltrs/tnandt1208.pdf.

MULTIAGENCY COORDINATION AT THE ATLAS TACK SUPERFUND SITE, FAIRHAVEN, MASSACHUSETTS

Agency coordination is an essential part of the Atlas Tack Superfund Site remediation. As part of planning for the ecological revitalization, EPA coordinated with the U.S. Army Corps of Engineers (USACE) and used the National Oceanic and Atmospheric Administration's (NOAA) Damage Assessment, Remediation, and Restoration Program (DARRP), which acts as a Federal natural resource trustee. NOAA contributed to the development of site-specific sediment remedial goals and the wetland removal plan, and greatly assisted in the design of the mitigation resulting in ecological revitalization at no additional cost to EPA. USACE and NOAA jointly designed separate fresh and salt water marshes to outcompete an invasive species at the site. Using remedial funding, three Federal agencies worked cooperatively to create an effective, natural remedy for the site. For more information, see Appendix A and visit www.epa.gov/ne/superfund/sites/atlas.

Trustees also conduct NRDAs, at sites with viable responsible parties, to calculate the monetary cost of restoring natural resources injured by releases of hazardous substances. They evaluate damages to natural resources by identifying the functions or "services" provided by the resources, determining the baseline level of the services provided by the injured resource(s), and quantifying the reduction in service levels because of the contamination. ERAs form the basis for establishing cleanup goals and may contain important information that EPA, Trustees, and risk assessors can use to evaluate ecological revitalization at a site.

Figure 2-1. Before and after photographs of the Atlas Tack Superfund Site in Massachusetts where the remedy resulted in preservation of wetland sediment and created a functioning wetland. See Appendix A for additional information. *Photographs courtesy of Elaine Stanley, EPA Region 1*

While property owners and communities generally conduct land use planning with input from stakeholders, it is important for EPA to understand the anticipated future uses for the site when planning and implementing the remedy. Establishing remediation goals for ecological receptors can be challenging if there is limited data on toxicity, effects on receptor species, and contaminant bioavailability. These challenges can be overcome by planning ahead and collecting appropriate ecotoxicological data (such as contaminant bioavailability and site-specific toxicity), reviewing the open literature and previous ERAs for data, and coordinating with stakeholders to identify site-specific receptors and past incidents of exposure. Uncertainties that cannot be addressed may be documented as part of the site-specific ERA and considered when selecting the site remedy or reuse. Stakeholders have the greatest reuse flexibility if remediation and reuse plans are coordinated *prior* to cleanup. EPA plays an important role in the planning process by communicating key information about the nature of contamination at the site, remedy options, and long-term protectiveness issues.

Stakeholders can still implement ecological revitalization even after the cleanup is complete. In 2004, EPA developed the Return to Use (RTU) Initiative to remove barriers to appropriate reuse at the hundreds of Superfund sites where cleanup has been completed. A focus of RTU has been on establishing partnerships with communities and other stakeholders to address potential obstacles to reuse. Through site-specific partnerships, referred to as demonstration projects, EPA is working with key stakeholders at RTU sites to identify potential reuse barriers and appropriate solutions for those obstacles (EPA 2008a). For more information on the RTU, visit www.epa.gov/superfund/programs/recycle/activities/rtu.html.

Coordinating Ecological Revitalization Efforts in the Superfund Removal Action Process. EPA has prepared a reuse assessment guidance for non-time critical removal actions (see Reuse Assessments Directive, OSWER 9355.7-06P, at *www.epa.gov/superfund/ programs/recycle/policy/reuse.html); however, guidance is not* currently available regarding reuse assessment for time-critical and emergency removal actions. The accelerated and time sensitive nature of these cleanups creates a challenge, as removal teams often complete their activities before there is an opportunity to consider reuse. In some cases, cleanup project managers can quickly conduct an ERA for a removal action, if there is an eminent threat to ecological receptors. However, these instances are rare and the removal action ERA follows the same process outlined for long-term ERAs conducted during the RI/FS. Because the time critical removal process is much faster than the remedial process, implementing reuse planning involves creating a targeted, expedited approach so that reuse can inform the removal action. For example, at the Calumet Container Superfund Site in Hammond, Indiana, EPA conducted a time critical removal action where ecological revitalization drove the reuse strategy for the site. In addition to contaminated soil removal, the removal action also included restoring wetlands and planting native plants. EPA worked successfully and expeditiously with stakeholders to determine future anticipated use of the site (see Appendix A for additional information about this site.)

Tools and Resources. The Superfund Program has developed and made available a variety of tools and resources supporting site reuse in general and ecological revitalization in particular (see *www.epa.gov/superfund/programs/recycle/tools/index.html* for a list of specific tools and resources that are available). In general, site managers can use SRI guidance documents to create and integrate reuse processes at sites undergoing either a remedial and removal action. SRI has also developed a community involvement process to advance reuse at remediation sites, which could be helpful at removal sites.

The Superfund Program has also developed several resources for site managers, consultants, and others interested in restoring disturbed sites. The Ecotools Web site (www.clu-in.org/ecotools) provides information on soil health, principles of ecological land reuse, and links to various federal, state, academic, and nonprofit agencies and organizations that support ecological revitalization. Through the Ecotools Web site, technical assistance is available for Superfund sites on various ecological revitalization topics, including ecological reuse of contaminated sites, use of soil amendments, use of native plants, control of invasive species, and re-vegetation. Fact sheets and Web-based seminars that focus on tools, methods, and technologies for implementing ecological reuse are also available. Answers to frequently asked questions related to ecological revitalization, re-vegetating landfills and waste containment areas, and attractive nuisance issues are available online at www.clu-in.org/pub1.cfm (EPA 2006c, d; EPA 2007c). The Green Remediation Web site (www.clu-in.org/greenremediation) provides various resources for cleanup project managers interested in incorporating green remediation strategies into cleanup actions. Resources include information on the use of BMPs; contracting and administrative toolkits; decision-making tools; links to initiatives involving green remediation applications; technical resources; and site-specific case studies. Technical assistance is also available for cleanup project managers in answering general inquiries about green remediation and for Superfund RPMs to build site-specific green remediation strategies. A useful resource available through this Web site is a technology primer on Green Remediation (EPA 2008j) that outlines the principles of green remediation and describes opportunities to reduce the carbon footprint of cleanup activities throughout the life of a project.

TECHNICAL ASSISTANCE FOR ECOLOGICAL REVITALIZATION AT SUPERFUND SITES

Regardless of the scope of the revitalization project, technical assistance can be obtained from the EPA's regional Biological Technical Assistance Groups (BTAG) (EPA 1991; see Appendix B for links to regional BTAG Web sites), EPA's Emergency Response Team (www.ert.org), EPA's Office of Superfund Remediation and Technology Innovation (OSRTI; www.epa.gov/tio), EPA's Ecotools Web site (www.clu-in.org/ecotools), and the U.S. Department of Agriculture's Natural Resources Conservation Service (www.nrcs.usda.gov).

In addition, groups such as regional Biological Technical Assistance Groups (BTAG), which are typically composed of biologists, ecologists, and ecotoxicologists from EPA, and agencies such as the U.S. Fish and Wildlife Service (USFWS), NOAA, and state environmental departments, could provide assistance during cleanup of a site to support ecological revitalization efforts.

2.3. Federal Facilities

EPA's FFRRO works with other EPA offices and federal entities to facilitate faster, more effective, and less costly cleanup and reuse of federal facilities. The federal facilities universe includes NPL sites and certain Base Realignment and Closure (BRAC) facilities (each subject to their respective provisions of CERCLA). The main difference between federal facilities and private Superfund sites is that at federal facilities, EPA has an oversight role rather than primary cleanup authority, which falls to the other federal agency. Many of the site-specific considerations for Superfund sites listed in Section 2.2 also apply to the federal facilities listed on the NPL as well as federal facilities not listed on the NPL (non-NPL sites). Additional challenges that might apply to federal facilities include special circumstances based on the contamination at that facility, such as munitions constituents.

FFRRO AND INTERAGENCY COORDINATION

In addition to EPA, FFRRO works with the following federal agencies to coordinate initiatives related to the cleanup of federal properties:

- Federal Aviation Administration
- Defense Logistics Agency
- National Aeronautics and Space Administration
- National Guard
- Small Business Administration
- U.S. Air Force
- U.S. Army
- U.S. Army Corps of Engineers
- U.S. Coast Guard
- U.S. Department of Agriculture
- U.S. Department of Defense
- U.S. Department of Energy
- U.S. Department of Interior
- U.S. Department of Transportation
- U.S. Navy

FFRRO's BRAC Program develops policies, plans, and initiatives to expedite the cleanup and reuse of closing military installations. Since 1993, the BRAC Program has worked with

U.S. Department of Defense (DoD), state environmental programs, local governments, and communities to achieve its goal of "making property environmentally acceptable for transfer, while protecting human health and the environment." For more information, visit the following Web site: www.epa.gov/fedfac/about ffrro.htm.

To implement congressionally mandated actions, EPA issued guidance on how to transfer federal facilities contaminated with hazardous wastes before cleanup completion. In the past, contaminated federal facilities had to undergo complete cleanup at least one year before transfer if hazardous waste was released from, disposed of, or stored on-site. Now, federal agencies can transfer properties prior to cleanup, as long they meet certain conditions. By transferring property that poses no unacceptable risks, communities benefit from faster reuse and redevelopment (EPA 2008c).

Midewin Tallgrass Prairie at the Joliet Army Ammunition Plant, Will County, Illinois

After working with the community and other stakeholders, the remediation team cleaned up contaminated soil through excavation and bioremediation. More than 19,000 acres of land was transferred to the Forest Service to create the Midewin Tallgrass Prairie, the first national tallgrass prairie in the country. While it will take years to fully restore the land, about a third is now open for the public to observe ongoing habitat restoration, as well as to hike, bike, or ride horseback on interim trails. For more detailed information about this example, see Appendix A.

Ecological revitalization is a part of many Department of Energy (DOE) and DoD facility reuse projects. Examples include Pease Air Force Base, JOAAP, Rocky Mountain Arsenal, Fernald, and Rocky Flats, which all have major ecological reuse components. See Appendix A for additional information on these case studies; the cover of this document includes a photograph of JOAAP.

Coordinating With Other EPA Offices and Programs. In carrying out its mission, FFRRO works closely with other EPA headquarters offices, including OSRTI, which manages the Superfund Program; ORCR, which manages the RCRA Corrective Action Program; and the Federal Facilities Enforcement Office (FFEO), which oversees compliance with environmental laws and guidance. EPA's Regional offices are also key partners in accomplishing EPA's federal facilities mission. RPMs and Community Involvement Coordinators (CICs), as well as toxicologists; attorneys; and reuse, tribal, and environmental justice coordinators based in each regional office work closely with EPA headquarters staff to coordinate site- specific cleanup activities. For issues requiring specialized expertise, FFRRO also collaborates with related EPA headquarters offices on a project-specific basis. Additionally, FFRRO co-chairs the Federal Facilities Leadership Counsel (FFLC), a coordinating body within EPA that provides direction and leadership on federal facility cleanup efforts. The FFLC is a forum for addressing a wide spectrum of federal facility cleanup issues, including compliance, technical, enforcement, financial, budgeting, and legislative issues. The FFLC includes EPA regional federal facility program and project managers, regional counsels, and headquarters staff from FFRRO and FFEO.

A Wildlife Refuge at the Rocky Mountain Arsenal in Commerce City, Colorado

EPA is partnering with the Army, Shell Oil, and the Colorado Department of Public Health and Environment to transform the Rocky Mountain Arsenal facility, one of the worst hazardous waste sites in the country, into one of the largest urban national wildlife refuges. The partnership is addressing contaminated ground water, surface water, soils, and buildings. Under the management of the U.S. Fish and Wildlife Service (USFWS), 27 square miles of open space surrounding the manufacturing facility is home to nearly 300 species of wildlife. After the cleanup is complete, the property will become a permanent part of the National Wildlife Refuge System (EPA 2008b). For more detailed information about this example, see Appendix A.

Coordinating With Other Agencies. FFRRO's partners include governmental and non-governmental groups that are involved in federal facilities cleanup. FFRRO works directly with other federal agencies, primarily DoD and DOE, to coordinate initiatives related to cleanup of federal properties.

FFRRO partners also include state, local, and tribal governments; community groups; environmental justice communities; and advocacy organizations. Local stakeholders include individuals, community groups and any other entities that might be affected by contamination, cleanup activities, or both. FFRRO encourages early and meaningful community involvement at all federal facilities.

Tools and Resources. FFRRO provides a variety of information resources about its programs, policies, and partners. The following Web sites provide access and information about its resources:

Visit www.epa.gov/fedfac/info.htm for access to EPA FFRRO's publications, newsletters, information centers, and other information resources.

Visit www.epa.gov/swerffrr/policy.htm for access to federal facilities related laws, regulations, policies, and guidance.

Visit FFRRO's comprehensive, searchable library of resources related to federal facility restoration and reuse topics at http://cfpub.epa.gov/ fdrl/index.cfm.

2.4. RCRA Corrective Action Facilities

EPA's ORCR regulates all household, industrial, and commercial solid and hazardous waste under RCRA, 1981, as amended. One important objective of EPA's RCRA Program is to protect the public from the management and disposal of hazardous wastes that RCRA facilities generate as part of normal operations. Examples of RCRA facilities include metal finishing operations, auto body repair shops, dry cleaners, chemical manufacturers, foundries, locomotive and railcar maintenance operations, and steelworks. In some cases, these facilities are no longer operational, have no significant activity, or are now vacant. Accidents or activities by hazardous waste generators or at hazardous waste treatment, storage, and disposal facilities regulated under RCRA may release contaminants into the environment. The RCRA Corrective Action Program ensures that regulated facilities that accidentally or otherwise release hazardous waste investigate and clean up such hazardous releases. The RCRA Corrective Action Program differs from Superfund in several ways. First, RCRA facilities often have viable owners and operators and on-going operations. As such, how best to use/reuse the property is ultimately the decision of the property owner, including whether to incorporate ecological revitalization elements on the facility. Second, EPA has delegated the RCRA Program to 43 states and territories that directly manage and oversee the Corrective Action Program; EPA implements the program in other unauthorized states.

BP Former Refinery, Casper, Wyoming

Under a RCRA Corrective Action Consent Decree, BP and the Wyoming Department of Environmental Quality (DEQ) cleaned up this 4,000-acre former refinery located along the banks of the North Platte River and incorporated several ecological revitalization components, creating wildlife habitat and allowing recreational reuse of the facility. Soda Lake, which was once used to dispose of waste water from the refinery, has been revitalized. BP worked with local citizens and the Audubon Society to design a bird sanctuary and resting ground for migrating birds. The reuse plan also incorporated a wetland treatment system into the design of a golf course constructed on the facility. The team planted more than 2,000 a trees as part of phytoremediation approach for cleaning up of portions of the property (EPA 2007a). This facility is a good example of how ecological revitalization measures can be incorporated at a facility with ongoing manufacturing activities. For more detailed information about this facility, see Appendix A.

In 1998, EPA established the RCRA Reuse and Brownfields Prevention Initiative to encourage the reuse of facilities subject to corrective action under RCRA so that contaminated or otherwise under-used land transitions back into productive use or greenspace (EPA 2008a). Several activities under this initiative support the ecological revitalization of RCRA facilities. One such activity is a cooperative agreement between EPA and the Wildlife Habitat Council (WHC). Under this agreement, the WHC works with EPA and other stakeholders to incorporate ecological revitalization into the cleanup design for end uses, hence providing wildlife habitat (WHC 2008). For example, corrective action at the Ford Rouge Center in Dearborn, Michigan, included ecological components to minimize impacts to the Rouge River. The cleanup team restored or created new wildlife habitat, including hedgerow wildlife corridors and wetland and grassland restoration. In addition to wildlife habitat, the project included other sustainable elements, such as installing a vegetated roof, using pervious pavement, and including phytoremediation. Because many aspects of the project involved ecological enhancement activities, the Ford Motor Company funded most of the activities on the property, with some additional funding provided through a state grant (for a stormwater swale) and an EPA grant to the Dearborn Public Schools System under its Five

Star Restoration Grants Program (to support wetlands restoration activities). See Appendix A for a case study regarding this facility.

DuPont-Remington Arms Facility, Lonoke, Arkansas

The DuPont-Remington Arms Facility continues to manufacture munitions on 385 acres of the 1,116-acre facility. The company manages the remaining 731 acres as a wildlife habitat. In cooperation with Ducks Unlimited, the cleanup team constructed a 20-acre moist soil impoundment for waterfowl habitat (EPA 2007b). See Appendix A for more detailed information about this facility.

EPA introduced RCRA Cleanup Reforms in 1999 (EPA 1999b) and additional Reforms in 2001 (EPA 2001) to more effectively meet the goals of the RCRA Corrective Action Program and speed up the pace of cleanups. One initiative of the 2001 Cleanup Reforms is capitalizing on the redevelopment potential of RCRA Corrective Action facilities. In addition, the RCRA program issued guidance to tailor cleanups to facility-specific end uses, including ecological end uses, while maintaining the ultimate goal of protecting human health and the environment. The "Guidance on Completion of Corrective Action Activities at RCRA Facilities" 68 FR 8757 (Feb 25, 2003) describes how corrective actions can be completed with contaminants remaining, using controls tailored to protection for a specific end use for the property (EPA 2005).

In most cases, facilities that are subject to RCRA corrective action continue their operations throughout the cleanup process. Although operations continue at these facilities, opportunities to incorporate ecological revitalization measures still may exist at parts of the property where there are no ongoing operations (see the DuPont-Remington Arms Facility text box). Facilities that are no longer continuing their current industrial or waste management operations may also provide opportunities for ecological revitalization. Some examples include the Ford Rouge Center in Michigan, the BP Oil facility in Lima, Ohio, and the Hopewell Plant (Honeywell) in Hopewell, Virginia. See Appendix A for additional information on these case studies. In some cases, especially with large properties, parcels of the property may provide special reuse opportunities (for example, riverfront location, road or rail access, or community reuse interest). In particular, many large RCRA facilities are federal facilities that may include large tracts of land that could be suitable for ecological revitalization or conservation easements. Stakeholders may be able to reuse uncontaminated parcels or those parcels on a shorter cleanup schedule more quickly than the entire facility (EPA 2008e). For example, at the former England Air Force Base in Alexandria, Louisiana, areas excavated as part of a remedial action became part of the Audubon Trail, providing habitat and a stopping point for migratory birds (see **Figure 2-2**). See Appendix A for additional information on this case study.

REUSE AT RCRA CORRECTIVE ACTION FACILITIES

In Spring 2001, a survey to determine trends in reuse potential of the 155 RCRA federal lead corrective action facilities in EPA Region 5 identified that 32 percent of all facilities (a total of 49) have potential for habitat or natural area restoration as a sole option or in combination with other reuses (EPA 2002b). While current, nationwide data is not available for ecological reuse of RCRA facilities, at least two regions (EPA Regions 3 and 10) recently conducted studies regarding their RCRA facilities' status and type of use. The results show that, even though most land use on RCRA facilities is industrial, as stakeholders reuse more RCRA facilities, a broader range of use is occurring. Visit the following Web site to review the results from EPA Region 3's study: www.epa.gov/region03/revitalization/ R3_land_ use_final/data_results.pdf.

Figure 2-2. Before and after photographs of England Air Force Base in Louisiana where contaminated areas were excavated and became part of the Audubon Trail, providing habitat and a stopping point for migratory birds. See Appendix A for additional information. *Photographs courtesy of RCRA Corrective Action Program*

Tools and Resources. ORCR provides a variety of information resources about its programs, policies, and partners. The following Web sites provide access and information about its resources:

Visit www.epa.gov/epawaste/hazard/correctiveaction/bfields.htm for information on the RCRA Brownfields Prevention Initiative and case study examples of successes under the initiative.

Visit www.epa.gov/epawaste/hazard/correctiveaction/resources for guidance and other information about RCRA corrective action.

2.5. Brownfields Properties

EPA's OBLR manages the Brownfields Program under the authority of Small Business Liability Relief and Brownfields Revitalization Act of 2002 (the "Brownfields Law"). EPA designed its Brownfields Program to empower states, communities, and other stakeholders to work together in a timely manner to prevent, assess, safely clean up, and sustainably reuse brownfields properties.

Brownfields are real property[1], the expansion, redevelopment, or reuse of which may be complicated by the presence or potential presence of a hazardous substance, pollutant, or contaminant. Included in the definition of Brownfields properties are sites contaminated with petroleum that represent a relatively low risk, including properties where the contamination resulted from an UST (Section 2.6 provides information on EPA's UST Program). An estimated 450,000 brownfields properties are located throughout the country (*www.epa.gov/brownfields/about.htm)*. Cleaning up and reinvesting in these properties relieves development pressures on undeveloped, open land while both improving and protecting the environment.

The Brownfields Program is a grant-based program that promotes green, ecological, and open space uses as part of its competitive grants process. These grants support revitalization efforts by funding environmental assessment, cleanup, and job training activities.

Brownfields funds can support sustainable remediation measures and planning for ecological revitalization (as the reuse of the property), but typically not actual revitalization or reuse activities. EPA's grant review process generally favors grant proposals that include ecological reuse as part or all of the ultimate reuse goals, especially with respect to greenspace and sustainable use criteria. The ultimate decision on whether a brownfields property will include ecological revitalization remains with the community receiving the grant. Although data specifically on the ecological revitalization of brownfields properties are not available, data reported by grantees on reuse measures for OBLR from fiscal year (FY) 2003 to FY2007 indicated that an estimated 4,756 acres were ready for reuse, and more than 507 acres of greenspace or open space were created (EPA 2008i). The Grace Lease property in Pennsylvania (see **Figure 2-3**) is an example of a restored Brownfields property, which had been dormant for nearly a century and was then converted into a natural habitat. A Brownfields Assessment Grant allowed stakeholders to study contaminant levels at the blighted property, remove uncertainties associated with property contamination, and transform the dormant property into usable greenspace for the community.

SEQ UIM BAY ESTUARY, JAMESTOWN S'KLALLAM TRIBE, WASHINGTON

The Jamestown S'Klallam Tribe used an EPA Brownfields Cleanup grant to clean up and restore estuary function to 82 acres of Sequim Bay. Cleanup activities included removing pilings, contaminated soil, and solid waste from the shoreline and riparian wetlands. The bay now provides clean sediment and habitat for shellfish, salmon, and other species. See Appendix A for more detailed information about this case study.

Figure 2-3. Before and after photographs of the Grace Lease Property in Pennsylvania, where a former industrial area was revitalized to natural habitat. See Appendix A for additional information. *Photographs obtained courtesy of Office of Brown fields and Land Revitalization*

Brownfields and Land Revitalization Technology Support Center (BTSC)

Coordinated through EPA's Technology Innovation Program, the BTSC ensures that Brownfields decision makers are aware of the full range of technologies available to make informed or "smart" technology decisions for their properties, including support for ecological revitalization. BTSC provides a readily accessible resource for unbiased assessments and supporting information on options relevant to specific properties, including a technology-oriented review process for investigation and clean-up plans for these properties. The BTSC also provides information about other available support activities, such as those conducted by the Technical Assistance to Brownfields (TAB) Program located at five regional Hazardous Substance Research Centers. Direct support is available to EPA regional staff, state staff, and local governments. For more information, visit www.brownfieldstsc.org.

The Brownfields Program also encourages the incorporation of green infrastructure into brownfields redevelopment projects. Green infrastructure techniques, such as bioswales, green roofs, and rain gardens, present an opportunity to return land to functioning and sustainable habitat. Other green infrastructure practices can also retain, treat, and release stormwater without exposing it to contaminated soils. For more information about this effort, visit www.epa.gov/brownfields/publications/swdp0408.pdf.

The Brownfields Program also provides Training, Research, and Technical Assistance Grants to fund projects that explore innovative ideas in the areas of protection of human health and the environment, sustainable development, and equitable development. Each assistance project will receive between $100,000 and $150,000 in annual funding for up to five years. Recipients can use the grants to support a variety of projects including, ecological revitalization, sustainable uses of land, and green jobs in communities. For more information about these grants, visit www.epa.gov/brownfields/trta.htm.

Other initiatives under the Brownfields Program can also contribute to ecological revitalization of brownfields properties. For example, through its partnership with Groundwork USA and the National Park Service Rivers, Trails, and Conservation Assistance Program, OBLR works with communities to improve their environment, economy, and quality of life through local action. This partnership also results in the ecological reuse of brownfields properties through Groundwork Trusts. Visit www.groundworkusa. net/index.html for more information about the Groundwork USA network.

Under the Sustainable Sites Initiative, EPA is currently working with the U.S. Green Building Council to provide a framework for the green development of brownfields properties. The framework is similar to what the Leadership in Energy and Environmental Design (LEED) system has accomplished for green buildings. The framework includes considerations for cleaning or mitigating all hazardous substances from prior use, supporting sustainable landscape principles and practices, and preventing the creation of future brownfields. For more information, see the following document: www. sustainablesites.org/report/SSI Guidelines Draft 2008.pdf.

Tools and Resources. OBLR provides a variety of information resources about its programs, policies, and partners. The following Web sites provide access and information about these resources:

Visit www.brownfieldstsc.org for information on strategies, technologies, and technical assistance available to support the investigation and cleanup of brownfields properties.

Visit www.epa.gov/swerosps/bf/toolsandtech.htm for access to a variety of tools and technical resources available to support property reuse.

Visit www.epa.gov/swerosps/bf/initiatives.htm for information on the various EPA and related initiatives that may be applicable at brownfields properties.

Visit www.epa.gov/swerosps/bf/partnr.htm to learn more about the partnerships that EPA has entered in support of brownfields revitalization and reuse.

2.6. Underground Storage Tank Sites

EPA's OUST manages and oversees the UST Program, which seeks to prevent leaks or releases of petroleum or certain hazardous substances from USTs, and ensures that contamination from USTs is cleaned up. OUST manages the program under the authority of several statutes, including Subtitle I of RCRA, as amended by the 1984 Hazardous and Solid Waste Amendments, the 1986 Superfund Amendments and Reauthorization Act, and the Energy Policy Act of 2005. States and territories primarily implement the UST Program, while EPA implements the UST Program in Indian Country. OUST administers the Leaking UST Trust Fund, which provides money for (1) overseeing and enforcing corrective action taken by a responsible party, who is the owner or operator of the leaking UST; and (2) implementing cleanups at UST sites where the owner or operator is unknown, unwilling, or unable to respond, or which need emergency action.

A key provision of the 2002 Brownfields Law allocates 25 percent of funding each year to assess, cleanup, and make ready for reuse petroleum brownfields properties that are relatively low risk. Of the estimated 450,000 brownfields properties in the U.S., approximately half are affected by USTs or some type of petroleum contamination (EPA 2008f). OUST is responsible for promoting the cleanup of sites with leaking USTs and coordinates with OBLR to refine the implementation of the law's petroleum provisions to allow more sites to support appropriate reuse or revitalization (EPA 2008d).

Pocket Park at a Former Service Station, Chicago, Illinois

A former service station in Chicago was transformed into a small pocket park using native plantings. This pocket park initiative is a joint effort by BP, the City of Chicago, and the local community. The contaminants of concern at the site were benzene, toluene, xylenes, and ethylbenzene (BTEX) at levels above maximum contaminant levels (MCLs) but not at levels that would pose a risk to the surrounding community. Once the site received "no further remediation" letters and was considered cleaned up, the team planted native species to create pockets of habitat for wildlife, expand greenspace for the community, and reduce stormwater runoff by reducing paved surfaces. See Appendix A for more detailed information about this example; this document's cover also includes a photograph of this pocket park.

To encourage the reuse of abandoned properties contaminated with petroleum from USTs, OUST created the USTfields Initiative in 2000. USTfields are abandoned or underused industrial and commercial properties where revitalization is complicated by real or perceived environmental contamination from USTs. The purpose of these pilots was to promote the importance of public- private partnerships; the critical role of the state as the primary implementing agency; and the leveraging of private funds to maximize cleanups. Although OUST will not award any new USTfields pilots beyond the original 50 pilots, sites may receive funding for similar assessment and cleanup projects through the Brownfields assessment, cleanup, and revolving loan fund grants and through the Leaking Underground Storage Tanks (LUST) Trust Fund.

Coordinating with Other Agencies. A major component of OUST's efforts to support the revitalization of contaminated sites caused by leaking USTs is collaboration with federal, state, and local agencies, and tribal and private partners to foster the revitalization and reuse of petroleum-contaminated sites. OUST also works with numerous grant recipients to enhance their efforts to revitalize petroleum brownfields. For example, OUST collaborated with the Indiana Brownfields Trails and Parks Initiative, which uses EPA grant funding to provide environmental assessments to local governments and nonprofits for brownfields properties (including petroleum brownfields) where parks, trails, or other green uses are planned (see www.in.gov/ifa/brownfields/files/TPI for more information on this state program). OUST is also partnering with EPA's Office of Policy, Economics, and Innovation (OPEI) to utilize several assistance mechanisms, such as the SmartGrowth America National Vacant Properties campaign. This campaign provides local planners with the information needed to consider viable reuse options, such as green or open spaces, at abandoned or under-utilized service stations and other petroleum brownfields.

OUST entered into a cooperative agreement with the WHC to help maximize the ecological benefits of reusing petroleum brownfields. One goal of the agreement is to demonstrate how federal, state, and local governments, tribal partners, industry, and community groups can use ecological revitalization to facilitate the restoration of petroleum brownfields for a variety of uses, including wildlife habitat. Under the agreement, the WHC will demonstrate the use of the latest technologies for applying ecological enhancements to site cleanups. Specific objectives for the partnership include: (1) achieving greater regulatory flexibility and support for ecological enhancements; (2) developing a strategy for obtaining constructive and meaningful stakeholder involvement; (3) ensuring sound scientific and technical support for ecological enhancement practices; and (4) promoting the value of ecological enhancements through a broad range of communication tools. OUST works with the WHC to identify opportunities to include ecological enhancements in end use plans at petroleum-contaminated sites. The pocket park project highlighted in the text box on the previous page is one of several successes resulting from this collaboration. WHC documents and provides case studies on a variety of programs on the following WHC Web site: www.wildlifehc.org/brownfield restoration/lust pilots.cfm.

OUST collaborated across all levels of government and with private industry to develop a Petroleum Brownfields Action Plan that improves stakeholder communications; expands technical assistance to states, tribes, and local governments; explores potential policy changes; and builds upon existing successes by expanding partnerships and testing new and innovative approaches to petroleum brownfields revitalization (EPA 2008d). The Action Plan provides a comprehensive framework for enhancing revitalization efforts at petroleum brownfields and promoting information sharing from both public and private sector efforts to revitalize petroleum brownfields. Four initiatives outlined in the Action Plan cover broad areas and can further EPA's collective efforts to highlight all applicable reuse options. Tasks within three of those initiatives are applicable to ecological revitalization and include the following:

- **Action Item 1.3** provides a basis for developing a "petroleum reuse/options catalogue" that could help compile and update information on reuse options and associated partnerships, as well as provide insights for interested parties to consider when addressing comparable sites.
- **Action Item 2.3** provides a framework to help eligible entities develop voluntary inventories of petroleum brownfields that complement local end use planning efforts.
- **Action Item 4.2** promotes the use of greenspace or wildlife habitat through collaboration with wildlife habitat organizations and property owners (of abandoned oil fields or urban petroleum brownfields) to support converting these properties to wildlife habitats.

OUST does not currently track the indicators listed in **Table 1-1** related to the status and type of end use. However, OUST is committed to tracking the mandatory measures and has developed the OUST Cross- Program Measures commitment memorandum (EPA 2007e). Petroleum brownfields sites are difficult to track and coordinate because of their small size, scattered distribution, variable ownership, and associated uncertainties in cleanup costs and liability. Continued coordination with organizations, such as the WHC, could help to provide a consistent means of tracking site reuse. Revitalizing petroleum sites also remains a local endeavor, and by enhancing public-private coordination, OUST intends to promote the appropriate use of petroleum brownfields sites to help meet community, end user, and stakeholder needs. Ultimately, though, local organizations drive the end use of each site.

Tools and Resources. OUST provides a variety of information resources about its programs, policies, and partners. The following Web sites provide access and information about its resources:

Visit www.epa.gov/swerust1/pubs/index.htm for publications that support the investigation and cleanup of leaking USTs.

Visit www.epa.gov/swerust1/rags/ustfield.htm to learn more about the USTFields Initiative and to access case studies on the pilot projects for examples and lessons learned associated with the reuse of former UST properties.

More information about the issues and opportunities associated with petroleum or UST brownfields cleanups is also available at *www.nemw.org/petroleum%20issue%20opportunity %20brief.pdf*(Northeast-Midwest Institute 2007; EPA 2008e).

3.0. Technical Considerations for Ecological Revitalization

There are several technical considerations for implementing ecological revitalization while cleaning up a property that are common to each of the cleanup programs discussed in Section 2.0. The objectives of ecological revitalization and those of the cleanup process are best accomplished if they are coordinated carefully. This section summarizes technical considerations for common cleanup and revitalization technologies that stakeholders can use during planning and design with the intent to minimize ecological damage during cleanups. Specifically:

- Section 3.1 presents factors to consider when selecting cleanup technologies for ecological revitalization.
- Section 3.2 addresses issues that may occur when waste is left in place at a cleanup property, how they could affect ecological revitalization, and potential approaches to mitigate these issues.
- Section 3.3 identifies ways to minimize ecological disruptions during cleanups.

3.1. Considerations When Selecting Cleanup Technologies for Ecological Revitalization

When designing and implementing any cleanup action at a contaminated property, it is necessary to consider certain factors related to natural resources or ecological revitalization (see text box below). Numerous *in situ* cleanup technologies can be used to ensure that contaminated properties are managed in a manner that protects human health and the environment; complies with federal, state, and local cleanup requirements; and allows for safe ecological revitalization. These cleanup technologies can include source control treatment (for example, soil vapor extraction and bioremediation), source control containment (for example, caps and barriers), institutional controls, and monitored natural attenuation. For additional information on a variety of cleanup technologies, visit EPA's CLU-IN Web site (www.cluin.org/techfocus) and the Annual Status Report (www.clu-in.org/asr). These cleanup technologies can affect ecosystems such as wetlands, streams, and upland areas such as meadows, prairies, and woodlands; therefore, it is important to consider their possible effects during ecological revitalization. While many of these effects are technology and property-specific, some general considerations apply, including the following:

- **Amendments:** Some *in situ* treatments involve adding amendments to the contaminated media. Project managers could evaluate their effects in the subsurface, their potential for eventual transport to surface waters, and their possible subsequent adverse effects on plant and animal communities. Some examples of soil amendments include organic matter additions such as biosolids, compost, manures, digestates, pulp sludges, yard wastes, and ethanol production byproducts; lime; wood ash; coal combustion products; foundry sands; steel slag; dredged materials; and water treatment residuals. At the California Gulch Superfund Site in Colorado, the remediation team applied lime and municipal biosolids to reduce the acidity of mine tailings and to reduce the bioavailability of heavy metals at the site (see **Figure 3-1**). For additional information on soil amendments, see the following document: www.cluin.org/download/remed/epa-542-r-07-013.pdf.

- **Regulatory requirements:** Federal and state regulations may apply to organic amendments such as biosolids, manures, and pulp sludges. State and local regulations apply to pH-adjusting amendments such as lime and wood ash as well as mineral amendments, such as foundry sand and dredged materials. For additional information, see the following document: www.cluin.org/download/remed/epa-542-r-07-013.pdf (EPA 2007d).
- **Attractive nuisance:** An attractive nuisance is an area, habitat, or feature that is attractive to wildlife, where waste or contaminants that have been left on site after a property is cleaned up that may be harmful to plants or animals. One objective of cleaning up such a property is to remove the pathway from a contaminant to a receptor. Some cleanup technologies, such as amended covers, are designed to prevent contact exposure, but they are not a barrier against burrowing animals. Preventing burrowing animals that could cause damage to a cleanup technology from entering the area, through fencing or other means, would help to keep the remedy intact, and protect the animals from coming in contact with the waste left on site. For additional information, see the following document: *www.clu-in.org/s.focus/c/pub/i/1438.*
- **Equipment and utility location:** Equipment generally needs periodic maintenance and monitoring. The cleanup team can maximize potential for habitat formation and biodiversity, and minimize disruption, by carefully considering the location of equipment. This might mean placing equipment near the edge, rather than in the middle, of a valuable habitat. For example, confining property disturbance to areas within 15 feet of roadways.
- **Hydrology and surface water management:** Cleanup technologies that could affect hydrology need to be designed carefully to avoid adverse effects on existing and anticipated habitat. For example, over pumping by ground water pump and treat (P&T) systems can cause dewatering of wetlands because over pumping lowers the water table (EPA 1993). Alternatively, discharging process water to surface waters and wetlands changes water depth, turbidity, circulation, and temperature. The use of settling basins and other such measures can help moderate discharges to wetlands and streams.
- **Surface vegetation:** Cleanup project managers are encouraged to consult technical experts to determine appropriate surface vegetation that will thrive but not interfere with the cleanup. For example, revegetation designed to emulate the native plant communities in the surrounding area would increase chances of success. However, vegetation growing near equipment related to a cleanup technology, such as a diversion wall, may prevent access to the equipment for maintenance and could cause performance issues. In addition, it is important to consider ecological succession when determining appropriate vegetation. Plant communities will naturally shift toward a climax community unless periodic maintenance is performed. When the cleanup technology, such as phytoremediation, employs vegetation, the plants selected to phytoremediate can also serve as a buffer to control runoff or

stabilize soil or streambanks. Stakeholders can obtain technical assistance through a variety of sources, including EPA's regional BTAG (www.epa.gov/oswer/risk assessment/ EPA's Emergency Response Team (www.ert.org), and EPA's Ecotools Web site (www.cluin.org/ecotools).

The considerations mentioned above, in addition to others shown in **Table 3-1** at the end of this section, play a role in addressing cleanup planning and design issues when considering ecological revitalization at properties where waste is left in place.

When Designing and Implementing a Cleanup Action, it is Important to Consider the Following:

- Physical and biological condition of the property and its location in relation to local and regional plant and animal species
- Regulatory requirements governing cleanup and protection or creation of ecologically significant areas
- Temporary and long-term ecological impacts
- Types of habitats that are to be protected, restored, or created at the property

Figure 3-1. Before and after photographs of the California Gulch Superfund Site in Colorado where site managers used high rates of lime amendment to neutralize the acidity of the mine tailings and applied municipal biosolids directly into the tailings along the Upper Arkansas River. See Appendix A for additional information. *Photographs courtesy of Michael Holmes, EPA Region 8*

3.2. Cleanup Planning and Design Issues and Ecological Revitalization

The text box at the right outlines some general steps when planning and carrying out ecological revitalization projects during cleanup planning and implementation. However, a number of issues associated with the application of a cleanup technology can alter the effectiveness of the cleanup or the ecological revitalization of a property. **Table 3-1** at the end of this section presents several issues that may occur when waste is left in place at a cleanup property, how they could affect ecological revitalization, and potential approaches to mitigate these issues. By carefully accounting for these issues at the outset, cleanup project managers can ensure the long-term success of the cleanup and minimize the potential negative effects of the cleanup approach on future uses of the property.

General Steps When Planning and Implementing an Ecological Revitalization Project

- Determine pre-disturbance and reference conditions
- Conduct a property inventory
- Establish revitalization goals and objectives
- Evaluate revitalization alternatives
- Develop a property-specific ecological design
- Prepare specifications for construction contractors
- Construct habitat features
- Conduct maintenance and monitoring activities

3.3. Minimizing Ecological Damage during Cleanups

Cleanups that include excavation and require earthmoving equipment can disrupt the surface area of a property and cause considerable loss of existing habitat as well as erosion, sedimentation, and colonization by invasive plants. These disruptions may also cause sedimentation or otherwise adversely affect ground water and nearby surface waters. To minimize the effects on habitat and encourage successful ecological revitalization, cleanup project managers may take steps to minimize excavation and other surface disruptions, avoid erosion and sedimentation, and protect the existing flora and fauna, by considering the following approaches (EPA 1993; Natural Resources Council [NRC] 1992; Kent 1994):

Develop and Communicate Ecology Awareness and Procedures. The process of ecological revitalization begins in the assessment or investigation phase, not after the remedy has been designed and is underway. Contractors and construction engineers are often not cognizant of sensitive ecological areas or aware that they can minimize disturbance and protect the ecology. Cleanup project managers can articulate a preservation policy and distribute it to everyone involved with on-site activities. Cleanup project managers can also incorporate requirements to protect habitat or species into construction plans, specifications, and contracts, as appropriate.

Design a Property-Wide Work Zone and Traffic Plan. The cleanup project manager can delineate staging areas, work zones, and traffic patterns to minimize unnecessary disruption of sensitive areas and existing habitat on or near a property. The cleanup team can delineate areas not requiring surface disruption and areas off-limits to disturbance, such as steep slopes, sensitive habitats, and clean stream corridors, with fences, tape, or signs to avoid disturbance by property workers and equipment.

Minimize Excavation and Retain Existing Vegetation. Earthmoving can destroy the roots of trees and other plants as well as disturb vegetation in uncontaminated areas. In addition, compaction of soil is also damaging to roots. These activities can be restricted to areas essential for the cleanup and avoided in all other areas. Some areas with low contamination levels or immobile contaminants posing no unacceptable risk to human health or the environment may be better off left undisturbed, if the disruptive effects of excavation outweigh the benefits of further cleanup, especially in valuable habitats (EPA 1998). Treatment and monitoring technologies are less invasive cleanup measures than excavation.

MYERS PROPERTY SUPERFUND SITE, NEW JERSEY

At the Myers Property Superfund site in Hunterdon County, New Jersey, (see case study in Appendix A), RPMs are saving select trees in areas with low levels of contamination by hand digging around the roots to a level of six inches. Excavated soil will be replaced with clean topsoil from off site. The site will be monitored in case large trees fall and expose soils deeper than six inches.

Phase Site Work. Sometimes cleanup project managers can phase construction by stabilizing one area of the property before disturbing another. This approach can reduce total soil erosion for the entire property and allows for revegetation or redevelopment of some areas immediately after cleanup. The cleanup project manager can also schedule construction to minimize the area of soil exposed during periods of heavy or frequent rains, and avoid sensitive periods (breeding, nesting, etc.) of certain species. For example, project managers at the Rocky Mountain Arsenal site (see case study in Appendix A and a photograph on the cover of this document) suspended cleanup activities during certain seasons to avoid disturbing the nesting and breeding of the bald eagle and other sensitive species.

Consider Property Characteristics. During the ecological revitalization of a property and to increase chances of successful revitalization, it is important that ecologists consider the following property characteristics: property size, existing habitat, proximity to undisturbed areas, topography, natural water supply, access, biodiversity (preserved by establishing connections between habitats or enlarging habitats), contaminant bioaccumulation (assessed during an ERA [EPA 1998, 1999a]), health of species and ecosystems, and threatened and endangered species (usually involves the assistance of a professional biologist or ecologist). Consider surrounding habitat when selecting native species for revegetation to increase chances of success. Urban properties pose additional challenges because they are typically small and may be subject to heavy runoff containing pollutants.

ROCKY MOUNTAIN ARSENAL, COLORADO

At the Rocky Mountain Arsenal, project managers recognized that cleanup-related traffic and road building could have major effects on the existing habitat at the 27- square-mile property. To facilitate reuse of the property as a wildlife refuge, they developed a property-wide traffic plan that routed traffic around valuable habitat and sensitive areas, minimized the potential for erosion and sedimentation, and used existing roads wherever possible. See the Rocky Mountain Arsenal case study in Appendix A for additional details.

Protect On-Site Fauna. In some cases, the project team may temporarily relocate on-site fauna that is being protected. Relocation may involve humane trapping and release, but less disruptive techniques may also be effective. For example, to relocate beavers and alligators at the French Limited Superfund Site in Crosby, Texas (see case study in Appendix A), project managers reduced their food supply in areas to be treated and increased the food supply in other suitable areas of the property. To protect fauna such as snakes, turtles, and some nesting birds that prefer edge habitat, it is necessary to consider careful use and parking of construction equipment in sensitive areas. For example, using construction equipment on edge habitat, or even using it to store equipment or fill material can adversely affect these species.

Locate and Manage Waste and Soil Piles to Minimize Erosion. Property cleanup may include the creation of temporary waste or soil piles to store contaminated soil for treatment or to store treated soil before redeposition. To minimize disruption of the local habitat, the cleanup project manager can structure stockpiles to minimize runoff; locate them away from steep slopes, wetlands, streams, or other sensitive areas; place them away from tree root zones to avoid soil compaction; and cover or stabilize them to control erosion and dust.

Design Containment Systems with Habitat Considerations. Building containment systems usually removes existing biota but can greatly improve the habitat, especially if the contamination present has severely degraded the area. While revegetation over containment areas or treatment systems must not detract from the effectiveness of the cleanup, cleanup project managers can design the cleanup components with ecological revitalization in mind. Cleanup project managers may also want to consider the type of contaminants, their stability, the media through which they travel, and the anticipated future land use. In addition, they may choose to avoid features that could damage the containment system or create an attractive nuisance. Where feasible, plan to allow enough soil above the protective cover to support the root systems of the intended vegetation. The use of fencing, removing access to potential food sources, or providing sufficient soil cover over the contaminated material can discourage wildlife from coming into contact with the contaminated material or from damaging a containment area.

Reuse Indigenous Materials Whenever Practical. Reusing logs, rocks, brush, or other materials found on site can provide logistical and ecological advantages as well as cost savings. Topsoil from on- site sources is usually well suited to support native vegetation. Treated soil and other materials can also be used as backfill, reducing the need for borrow areas for clean fill. Green waste, such as logs and branches can be used on site, to a limited degree, to create structure within the new habitats. Excess woody material can be shredded, composted, and used as a soil amendment. For example, at Loring Air Force Base in Northeastern Maine (see case study in Appendix A), boulders and cobbles, larger than 15 centimeters in diameter, were removed from the streambed and nearby trees during cleanup and later used in stream reconstruction, after completion of cleanup activities. Reuse of native materials at this property significantly reduced the need for additional materials and thereby achieved cost savings.

Control Erosion and Sedimentation. Revitalization areas usually need erosion and sedimentation control measures to avoid disturbing sensitive areas, even when state or local regulations do not require them. These measures can include retaining sediment on the property and managing runoff using filters, such as compost or other organic materials.

Ensure that Borrow Areas Minimize Impact on Habitat. Borrow areas, locations where cleanup teams excavate clean soil for use elsewhere during a cleanup, may be located and used with ecological revitalization objectives in mind. For example, borrow areas can be located in low-value areas to create or improve habitat and be designed, contoured, and vegetated to meet aesthetic and habitat considerations. Based on consultations with the USFWS, project managers at the Rocky Mountain Arsenal (see case study in Appendix A and a photograph on the cover of this document) designed borrow areas to establish the habitat of a planned wildlife refuge.

Avoid Introducing New Sources of Contamination. If not properly managed, cleanup activities can introduce new sources of contamination that may affect habitat and ecological receptors. Contamination can result from materials used on the property, fugitive dust emissions, and operations of equipment and sanitation facilities. Materials that can cause contamination include pesticides, herbicides, fertilizers, petroleum products, treatment agents, and solid wastes. To avoid introducing these new sources, storage areas can be sheltered from the elements, lined with plastic sheeting, surrounded by berms, and regularly inspected for releases. In addition, equipment maintenance can be done in suitable staging areas and adequate sanitation facilities for property workers can be provided away from streams, wetlands, and other sensitive areas.

Prevent the Introduction of Undesirable Species. Non-native plant species can invade and destroy native species. To prevent introducing undesirable species, monitor barren and disturbed areas, which are susceptible to colonization by undesirable plants, and remove undesirable species where necessary. In addition, equipment operators can wash trucks and equipment before entering a property to avoid introducing invasive plant seeds. Clothing and shoes can also be managed to avoid introducing invasive plant seeds

Table 3-1. Cleanup Planning and Design Issues When Waste is Left on Site and Other Considerations for Ecological Revitalization

Issue	Property Type[2]	Potential Impact	Solution/Consideration
Attractive Nuisance Issues: An area, habitat, or feature that is attractive to wildlife and has, or has the potential to have, waste or contaminants left on site that are harmful to plants or animals after a property is cleaned up	Landfill Mining Site Brownfield Military Installation Foundry Gas Station Metal Plating Facility Refinery Tannery	• Harm wildlife if (1) an exposure pathway exists from contaminants left on site that could directly harm wildlife or travel up the food chain; or (2) wildlife interfere with the cleanup, thereby creating an exposure pathway	• Consider potential ecological risks throughout the cleanup process • Conduct a thorough ecological risk assessment to avoid potential attractive nuisance issues • Carefully consider plant species and the type of animals that those species will attract; protect newly planted species until they are established • For additional information, refer to EPA's fact sheet titled "Ecological Revitalization and Attractive Nuisance Issues" (EPA 2007c)
Managing Gases: Depending on the waste composition, some containment sites have the potential to generate gas	Landfill	• Provide fuel for fire or explosions • Stress vegetation • Damage cover system • Infiltrate nests or other wildlife homes • Create other health or safety hazards	• Determine ability of waste to generate gas during planning stage (EPA 1991) • Build gas collection systems • Place components where they (1) do not interfere with planned uses, (2) minimize noise and odors, and (3) are not easily accessible to trespassers or wildlife • For additional information, refer to the EPA fact sheet "Reusing Cleaned Up Superfund Sites: Commercial Use Where Waste is Left On Site" (EPA 2002a) and "Landfill Gas Control Measures" (www.atsdr.cdc.gov/HAC/landfill/PDFs /Landfill_2001_ch5.pdf)

Table 3.1 (Continued)

Issue	Property Type[2]	Potential Impact	Solution/Consideration
Restoring Soil: Soils, especially those found in urban, industrial, mining, and other disturbed areas suffer from soil toxicity, too high or too low pH, lack of sufficient organic matter, reduced water-holding capacity, etc.	Mining Site Manufacturing Facility Metal Plating Facility Brownfield Refinery Tannery	• Decrease ability to support vegetation, which can lead to increased erosion and offsite movement of contaminants by wind and water	• Consider appropriate soil amendments (inorganic, organic, or a mixture) to limit contaminant bioavailability and restore appropriate soil conditions for plant growth by balancing pH, adding organic matter, restoring soil microbial activity, increasing moisture retention, and reducing compaction
Settlement: The consolidation of subsurface materials at closed-in-place sites due to compaction or degradation	Landfill	• Rate and magnitude of settlement may affect the type of habitats that will be successful • Damage containment systems, alter slopes, cause gullies to form, and disturb other property features • Municipal landfills can settle up to 30 percent of the landfill depth over 15 to 30 years	• Consult with geotechnical engineer during cleanup planning to estimate settlement magnitude, distribution, and rate • If necessary, delay ecological revitalization until settlement has largely ceased, but under long-term settlement scenarios, vegetation will likely adapt to the changing property conditions • Use a nurse crop like oats, to control erosion and provide greenspace • Use construction techniques, such as preloading, vibrocompaction, and dynamic compaction, to accelerate settlement (these approaches will not affect settlement caused by biodegradation); however, do not compact topsoil because over-compaction of topsoil will result in vegetative failure

Table 3.1 (Continued)

Issue	Property Type[2]	Potential Impact	Solution/Consideration
Stabilizing Metals: Some property soils contain toxic levels of metals that can be harmful to plants or animals	Mining Site Metal Plating Facility Brownfield Refinery Tannery	• Metals taken up by plants which are eaten by animals causing a potential attractive nuisance • Metals leach into ground water	• Use soil amendments to chemically precipitate or sequester metals that are present in the soil; this can reduce metal availability to plants and metal leaching into water • Select plant species based not only on availability but also on their ability to establish and grow in a newly created root zone and the species' inability to uptake metals
Surface Vegetation: Used to limit soil erosion, promote evapotranspiration and surface water management, and, in some cases, may be a component of the cleanup (for example, phytoremediation)	Landfill Mining Site Brownfield Military Installation Foundry Gas Station Metal Plating Facility Refinery Tannery	• Not all plants are well-suited to property conditions • Roots can physically damage equipment for a cleanup treatment technology, such as a barrier or well	• For wetlands, study the proper hydrology, tidal elevation, and height of a newly constructed wetland profile; these factors are of great importance to allow the new wetland (both saline and fresh) to flourish • When selecting plants, consider Executive Order (EO) 13148, which promotes use of native species • Place equipment away from areas where deep-rooted vegetation will be planted • Choose native plants found in the surrounding natural areas because they have the most chance of success, require the least maintenance, and are the most cost-effective in the long term • Ensure the waste containment system is properly designed and implemented to maintain system integrity while supporting a variety of plants • For additional information, refer to EPA's fact sheet titled "Revegetating

Table 3.1 (Continued)

Issue	Property Type[2]	Potential Impact	Solution/Consideration
			• Landfills and Waste Containment Areas Fact Sheet" (EPA 2006d)
Surface Water Management: Includes a variety of activities that protect the natural functions and beneficial uses of surface waters	Landfill Mining Site Brownfield Military Installation Foundry Gas Station Metal Plating Facility Refinery Tannery	• Affects nearby vegetation, streams, lakes, and wildlife migration routes through erosion or sedimentation • Runoff controls and water diversions implemented as part of a cleanup influence water tables and the rate of flow into streams or wetlands • Erodes the top layer of a cover system • Percolates into a cap	• Design protective caps to prevent precipitation from infiltrating into the subsurface and grade the cap to establish an effective slope (usually 3-5 percent) • Route runoff through settling basins to collect sediment to reduce impacts to property hydrology and construct runoff controls to reduce the volume and rate of runoff to low-lying areas, wetlands, or streams • Use rerouted runoff to create new wetland habitat or enhance existing habitat to provide natural controls and reduce contaminant transport • Build drainage channels and swales and design diversions where possible to minimize changes to natural drainage patterns or the quantity of surface water flows to wetlands or streams • For additional information, refer to EPA's fact sheet titled "Controlling the Impacts of Remediation Activities in or Around Wetlands" (EPA 1993)
Timing: The time at which ecological revitalization is considered during the remedial planning process	Landfill Mining Site Brownfield Military Installation	• The longer planning is delayed, the greater the possibility that fewer reuse options will be available	• Begin revitalization planning as early as possible • Begin developing a revitalization project on parts of a property before a cleanup is completed, if possible

Table 3.1 (Continued)

Issue	Property Type[2]	Potential Impact	Solution/Consideration
	Foundry Gas Station Metal Plating Facility Refinery Tannery		• Consider advice from a restoration ecologist to determine the proper season to plant grasses, shrubs, and trees • Consider breeding seasons and other timing issues to avoid affecting sensitive species when scheduling remedial or revitalization activities
Utilities: Can include sanitary sewers, water, telecommunications, natural gas, and electricity	Brownfield Landfill Manufacturing Facility Military Installation Foundry Gas Station Metal Plating Facility Refinery Tannery	• Act as a conduit for gas migration • Facilitate water infiltration into a waste containment area • Require excavation into a waste containment area and contaminated material if utility repairs are necessary • Increase the quantity of leachate generated if sewer lines below a waste containment area begin to leak • Can be damaged by settlement	• Include special provisions to ensure utilities do not hinder the effectiveness of the cleanup or ecosystem functions; for example, avoid burying a utility line in a protective cap or placing it in an area where trees will be planted • For additional information, refer to the following EPA report: "Reusing Cleaned Up Superfund Sites: Commercial Use Where Waste is Left On Site" (EPA 2002a)

[2] See Table 2-1 for EPA Programs that can apply to each property type.

4.0. Wetlands Cleanup and Restoration

Wetlands are of particular concern for cleanups because in addition to intercepting storm runoff and removing pollutants, they provide food, protection from predators, and other vital habitat factors for many of the nation's fish and wildlife species (EPA 2008g). Section 3.0 discusses the general considerations that apply during planning and design of a wetland cleanup and restoration. This section summarizes wetland cleanup and restoration, focusing on specific considerations during planning and design.

. Whether a cleanup involves restoring an existing wetland or creating a new one, a cleanup project manager must typically take the following steps (EPA 1988; USFWS 1984):

- Evaluate the characteristics, ecological functions, and condition of wetlands related to the property
- Determine the type of wetland functions and structures that would be beneficial in the area after the cleanup
- Develop a wetland design that will achieve the stated ecological functions
- Design the cleanup and wetland features to ensure that cleanup activities have minimum effect on existing wetlands and other ecosystems and do not create an attractive nuisance (see **Table 3-1** for additional information on attractive nuisance issues)
- Specify and implement maintenance requirements

Once it has been determined that a cleanup will affect a wetland, several key factors need to be considered, including the following:

Wetland Characteristics. The cleanup project manager may wish to determine wetland characteristics to develop a thorough understanding of the role of the wetland in the overall ecosystem and the relationships between the various plant and animal species within the wetland. It is also important to determine if any endangered, sensitive, or commercially important wetland species are present.

Wetland Regulatory Requirements. Several regulatory requirements generally apply when a cleanup or reuse project affects wetlands, including Sections 401, 402, 403, and 404 of the Clean Water Act; Section 10 of the Rivers and Harbors Appropriation Act; and the Federal Agriculture Improvement and Reform Act, commonly known as the Farm Bill. Depending on the type of cleanup and the law under which action is taken, permits may be needed prior to conducting any cleanup activities.

Wetland Mitigation and Ecological Revitalization

Cleanup project managers may consider ecological revitalization part of wetland mitigation depending on the property-specific habitat. However, if the wetland mitigation is part of a contaminant treatment system and is not intended to provide habitat, it cannot be considered ecological revitalization. For additional information on wetland mitigation requirements, go to www.epa.gov/wetlandsmitigation. For additional information on wetlands in general, go to www.epa.gov/wetlands.

Wetland Vegetation and Hydrology. Analyses of hydrologic and soil conditions help define the property's wetland vegetation associations (a known plant community type, uniform habitat conditions, and uniform appearance). Generally, restoring hydrology and reestablishing a previous vegetation association tends to lead to a successful wetland ecosystem. For properties where the historical native vegetation association cannot be determined, use nearby wetlands with similar soil and hydrology as a guide. See example in text box to the right and Figure 4-1 at the end of this section. For additional information on reference wetlands, visit the Society for Ecological Restoration's Web site under Section 5 of the Ecological Restoration Primer: www.ser.org/content/ecological_restoration_primer.asp. Also, consider water availability and soil type when selecting and placing the vegetation. Where appropriate, seeded species that establish quickly may be planted first, followed by species that are more difficult to establish. Where available, a natural seed bank in existing wetland soils is often adequate for establishing wetland vegetation.

Wetland Wildlife. Wetlands provide valuable wildlife habitat. The ability of a wildlife species to thrive in a wetland is dependent upon a number of factors, including the minimum habitat area necessary for the species, the minimum viable population of the species, the species' tolerance for disturbance (for example, excavation or installation of ground water pumps), and the wetland ecosystem's functional relationship to adjacent water resources and ecosystems. Thus, three factors will play a major role in determining the effectiveness of a wetland for long-term wildlife use: (1) the size of the wetland, (2) the relationship of the wetland to other wetlands, and (3) the level and type of disturbance (Kent 1994; NRC 1992; EPA 1994).

USE OF NEIGHBORING WETLANDS AS REFERENCE AT NAVAL AMPHIBIOUS BASE LITTLE CREEK, VIRGINIA BEACH, VIRGINIA

After removing a 1.2-acre landfill, the Navy, in partnership with EPA and Virginia Department of Environmental Quality, constructed a tidal wetland in the Chesapeake Bay. The team achieved tidal wetland hydrology by constructing two connecting channels to the nearby Little Creek Cove. In addition, they used a neighboring marsh as a reference wetland to determine appropriate plants to place along designated elevations to establish tidal wetland vegetation. See

Appendix A for additional information on this case study.

Wetland Maintenance. A variety of wetland maintenance activities are needed to ensure long-term success, including weed control and management of aggressive exotic species, such as common reed (*Phragmites australis*), purple loosestrife (*Lythrum salicaria*), water hyacinth (*Eichornia crassipes*), and salvinia (*Salvinia molesta*). In addition, installing wire screens or other barriers around the plants or the planted area to control deer, rabbit, or beaver grazing can help protect vegetation until the ecosystem becomes established. Periodic monitoring of the wetland for plant loss, erosion, insect or disease infestations, and litter or debris buildup is also important. For properties near populated areas, public education efforts can help reduce maintenance issues associated with litter or debris dumping, off-road vehicle use, or other human activities that may threaten the long-term success of a wetland project.

Figure 4-1.Before and after photographs of Naval Amphibious Base Little Creek in Virginia, where the remediation team converted a landfill into a tidal wetland. See Appendix A for additional information. *Photographs courtesy of Bruce Pluta, EPA Region 3*

Bunker Hill Superfund Site in the Coeur d'Alene River System in Kellogg, Idaho

At the West Page Swamp area of the Bunker Hill Superfund Site, EPA contractors spread a cap composed of compost and wood ash over the soil to reduce accessibility and bioavailability of the underlying tailings and to restore wetland function.

Treatment Wetlands. Wetlands created to treat contaminants have some additional considerations regarding ecological revitalization and attractive nuisance issues. Conducting an ERA and monitoring of the treatment wetland until it meets cleanup goals can help to identify any potential attractive nuisance issues. Cleanup project managers are employing this approach on a variety of cleanups. For example, a public-private partnership is installing a series of passive treatment systems, including treatment wetlands, to treat acid mine drainage from abandoned surface and underground coal mines in western Pennsylvania. After passing through a series of limestone-lined ponds to neutralize pH, the water is sent through an aerobic constructed wetland to remove iron hydroxides. The system can even recover metals removed from the water so recovered metal can be sold (see Appendix A for additional information on this case study).

Treatment wetlands are also used as the final polishing treatment step of a remediation scheme. For example, stormwater or effluent from ground water treatment systems can be sent through restored or created wetlands before being released to nearby waterways. This step helps remove suspended solids and other pollutants from the stormwater or effluent.

Ideally, cleanup goals will be met when using a treatment wetland to assist in property cleanup. Once the property meets its cleanup goals, components of the remedy, including a wetland, may no longer be necessary for further treatment. At this stage, coordinating with co-regulatory partners to determine long-term maintenance and stewardship responsibility for the wetland is critical. Section 7.0 discusses long-term stewardship.

For additional information on treatment wetlands, visit the following Web site: www.epa.gov/owow/wetlands/ watersheds/cwetlands.html.

5.0. Stream Cleanup and Restoration

Stream cleanup and restoration are important because streams serve as corridors for migratory birds and fish, and they provide habitat to many unique species of plants and animals (EPA 2008g). Cleaning up a stream corridor can be complicated, as cleanups often disrupt the stream flow and habitat. This section provides an overview of considerations for designing and implementing cleanups that facilitate ecological restoration of streams and stream corridors and mitigating adverse ecological impacts of constructing cleanup features. A successful stream cleanup, combined with appropriate restoration strategies can hasten the recovery of degraded stream corridors and begin the natural process of restoring their ecological functions (EPA 1995).

Importance of Stream Corridors

Healthy stream corridors can provide important habitat for fish populations; erosion and sedimentation control; high-quality water for wildlife, livestock, flora, and human consumption; opportunities for recreationists to fish, camp, picnic, and enjoy other outdoor activities; and support for diverse plant and wildlife species.

An important first step in cleaning up a stream corridor is to assess the possible sources of disturbance from cleanup activities. Baseline data can be gathered on existing species, in-stream and riparian habitat, soil characteristics, and stream function to characterize potential degradation. Other disturbances to characterize include stream channel alteration, water quality impairment, invasion by exotic species, loss of riparian vegetation, and compaction or undercutting of streambanks. Defining the conditions of the stream corridor prior to the disturbance can help to identify the cause of the disturbance. Another important step is to determine the type of ecosystem that can be established in the stream corridor. When historical records are unavailable, information on undisturbed, nearby stream corridors with similar physical characteristics can help determine the type of ecosystem that will likely be successful at the property. The following considerations are critical to a successful stream cleanup and restoration:

Stream Channel Restoration. Removing contaminated sediment and soil from stream channels and banks during a cleanup typically results in severe alteration of stream flow. In such instances, reconstruction of stream channels and banks is usually necessary. Decisions about stream channel width, depth, cross-section, slope, and alignment profoundly affect future hydrology (and the resulting ecology) of the stream system. Restoration design typically considers factors such as the physical aspects of the watershed, hydrology, sediment size distribution, average flood flows, and flood frequency. When designing a stream channel restoration, the cleanup project manager can try to anticipate the effects of future land uses on the watershed. For example, the restoration of riverbanks along the Poudre River was designed to accommodate heavy recreational use while providing ecological benefits (see case study in Appendix A). For additional information, refer to resources listed in Appendix B and the following publication at www.clu-in.org/download/ newsltrs/ tnandt1208.pdf.

TIDAL CHANNELS

Stream channel restoration can include tidal channels. After removing contaminated sediment at the Atlas Tack site in Fairhaven, Massachusetts, site managers used coconut coir fiber logs to stabilize the salt marsh tidal channels. See Appendix A for additional information on this case study.

Streambank Stabilization. Disturbed or reconstructed streambanks often need temporary stabilization to prevent erosion. Temporary stabilization can consist of natural materials such as logs, brush, and rocks, and property planners can design it so as not to hinder permanent revegetation. At the Cache La Poudre River Superfund Site, EPA incorporated boulders and snags into the cleanup to stabilize the streambank while providing habitat (see **Figure 5-1** and case study in Appendix A). In some cases, geotextiles, natural fabrics, and bioengineering techniques may be necessary. Revegetating streambanks using seeding or bare root planting techniques will often fail if the stream floods before vegetation is fully established. Consequently, temporary vegetation for stabilizing streambanks may be more successful using anchored cuttings or pole plantings (that is, woody cuttings or poles inserted and anchored into the streambank) taken from species that sprout readily, such as willows. For additional information, refer to resources listed in Appendix B.

FORT COLLINS STREAM CORRIDOR RESTORATION

In Fort Collins, Colorado, soil and ground water contamination migrated to the Cache La Poudre River and contaminated the sediments of this wild and scenic river. Cleanup activities included temporarily re-routing the river and excavating the contaminated sediments. The remediated portion of the river was not channelized, and EPA made an effort to create an unobtrusive remedy by consulting ecological restoration experts to create natural stream characteristics. See Appendix A for additional information on this case study.

Streambank Vegetation. Wherever possible, it is important to protect existing native vegetation, especially mature trees, during cleanup and restoration activities; however, many properties will need some revegetation. Cleanup project managers may select species for revegetation for their ability to establish a long-lasting plant community rather than as quick fixes for erosion or sedimentation problems. For example, fast growing non-native species may quickly stabilize a denuded stream bank, but over the long term, they may end up invading the entire stream corridor to the detriment of desirable native species. Approaches that attempt to establish ecosystems similar to pre-disturbance conditions tend to have more long-term success and need less maintenance than more highly engineered solutions (for example, gabions or riprap) that reduce the amount of viable habitat. For additional information, refer to resources listed in Appendix B.

Figure 5-1. Before and after photographs of the Cache La Poudre River Superfund Site in Colorado, where EPA implemented an ecological remedy to preserve the riverine habitat and restore the streambank. See Appendix A for additional information. *Photographs courtesy of Paul Peronard, EPA Region 8*

Watershed Management. The entire watershed ecosystem affects the health and condition of a water body. Therefore, cleanup and revitalization may need to address watershed processes that degrade ecosystems, such as sediment loading from road cuts or construction, increased runoff from impervious areas, and other point and nonpoint sources of pollution. Effective watershed management could even eliminate the need for in-stream restoration approaches.

Bioengineering techniques have become an increasingly popular approach to streambank restoration and maintenance. Bioengineering refers to stabilizing the soil or streambank by establishing sustainable plant communities. Stabilization techniques may include using a combination of live or dormant plant materials, sometimes in conjunction with other materials such as rocks, logs, brush, geotextiles, or natural fabrics. Bioengineering techniques can be more labor intensive than traditional engineering solutions and sometimes take longer to control streambank erosion. Nevertheless, over the long term, they often have lower maintenance costs and create important habitat.

Finally, maintenance such as erosion control, reseeding, and soil amendments may be needed after evaluating the initial progress of stream corridor recovery. Allowing natural processes to shape the ecosystem in the stream corridor will generally lead to self-sustaining, long-term recovery of in-stream, riparian, and upland terrestrial habitats in the stream corridor. Because this process takes time, providing short-term riparian and upland habitats may hasten the return of wildlife to the disturbed area. Cleanup project managers may use engineered habitat structures such as weirs, dikes, randomly placed rocks, riffles and pools, fish passage structures, and off-channel pools to enhance in-stream habitat during the short term. Engineered habitat structures are most effective when installed as a complement to a long-term recovery strategy. For additional information on engineered habitat structures, see Section 8G of the Federal Interagency Stream Restoration Working Group's Stream Corridor Restoration Guide at www.nrcs.usda.gov/Technical/stream restoration/newtofc.htm.

6.0. Terrestrial Ecosystems Cleanup and Revitalization

Grading or earthmoving operations at cleanup properties can seriously disturb terrestrial plant and animal life at properties. The cleanup process can denude some contaminated properties of all vegetation and topsoil. Establishing a plant community that will thrive with minimal maintenance is a critical step in developing a healthy terrestrial ecosystem on these properties. This section discusses factors to consider when planning terrestrial plant communities in disturbed areas. It addresses (1) general revegetation principles and factors to consider in the course of protecting or creating natural terrestrial ecosystems and (2) specific considerations when creating meadows or prairies and establishing vegetation on semi-arid or arid lands. Section 3.1 presents general cleanup planning and design issues that may also be applicable to the revitalization of terrestrial ecosystems.

General Revegetation Principles. While restoring terrestrial ecosystems, it is recommended that cleanup project managers consider soil type, plant selection, and timing.

Native Plantings at College Park Landfill

At the College Park Landfill in Beltsville, Maryland, cleanup project managers used recycled waste materials such as fly ash and animal and plant by-products as land cover as part of the landfill cap. In addition, the vegetative cover includes diverse native plantings. See Appendix A for additional case study information.

Soil Type. Soil testing is generally necessary to evaluate whether the pH, nutrient availability, toxicity, salinity, and organic material content are appropriate for successful plant establishment. Several organizations provide assistance in soil testing, including U.S. Department of Agriculture (USDA)'s Natural Resources Conservation Service (NRCS) and the WHC. The soil can then be prepared or amended, as necessary, to ensure proper soil texture and conditions. Soil amendments, or residuals from other processes that have beneficial properties when added to soil, may be used in areas without adequate topsoil; if fertilizer is needed, it is important to choose a formulation that meets the growing needs of the selected species (EPA 2007d). The cleanup team may also have to stabilize the soil and apply compost to hold seed in place, aid in establishing plants, mitigate the effect of rainfall on newly seeded areas, preserve soil moisture, and control erosion. Soil stabilization methods include mulching with straw or wood-fiber product, or installing synthetic matting. Cleanup project managers may wish to select soil amendments and stabilization techniques for their ability to improve conditions for germination of the selected species. In addition, some types of soil amendments may help adjust the pH of the soil in preparation for seeding (EPA

2007d). Refer to the following document for more information on soil testing: www.nrcs.usda.gov/feature/backyard/pdf/nutrient.pdf.

AMENDING SOILS WITH BIOSOLIDS AT A REFINERY

In Lima, Ohio, a refinery undergoing RCRA Corrective Action is using biosolids to help create prairie habitat with native grasses, flowers, and trees over a soil cover. See Appendix A for additional case study information.

Plant Selection. Seed mixtures and plants can be adjusted to suit the soil, climate, hydrology, exposure (to both sun and wind), and topography of an area. Local native populations of plant and seed usually result in higher survival rates and maintain the integrity of the local gene pool. As discussed in Section 3.0, cleanup project managers are encouraged to avoid using non-native species. These species can out-compete and displace native species, disrupt ecological processes, and significantly degrade entire plant communities, both on and off the property.

After seeding, cleanup project managers can protect the seeded areas from grazing animals, vehicles, and other disturbances until plants are well established. Techniques for protecting plantings include fencing, clearly marked access roads, animal repellants, trenches or berms to control run-on and runoff (if they are already part of stormwater control features at the cleanup property), and interim surface stabilization methods such as mulching or matting. Cleanup project managers may need to reseed the area within the planting season to replace damaged vegetation or to achieve the desired plant density. For additional information on seed mixtures and plant selection, visit EPA's GreenAcres Web site (*www.epa.gov/ greenacres),* the Plant Conservation Alliance (PCA) Web site (www.nps.gov/plants), and the Bureau of Land Management's Seeds of Success Program (www.nps.gov/plants/sos).

Timing. It is important to seed during the optimum periods for plant establishment, which are property- specific and vary depending on the type of terrestrial habitat that is being restored. Information on seeding techniques and conditions for individual species is available from NRCS technical guides (www.nrcs.usda.gov), university extension offices, and seed suppliers. If planting cannot occur during optimum periods, cleanup project managers may use a nurse crop, such as annual rye or oats, as ground cover until the appropriate planting season.

Meadows and Prairies. A few additional considerations apply when restoring meadows or prairies. Generally, when seeding an area with native grass species, specialized planting equipment, such as a native grass drill, is needed to ensure good seed to soil contact. Seeds need to be certified and purchased on a pure live seed basis. Grass stands usually do not need fertilizer or irrigation. However, they may need periodic maintenance activities, such as controlled burning, mowing, and removing plant litter, to suppress woody growth and encourage vigorous new growth. To maximize benefits to wildlife, conduct these activities outside of the primary nesting season, preferably in late winter or early spring.

Semi-Arid and Arid Areas. Cleanup project managers may consider a number of additional factors when establishing vegetation in semi-arid and arid areas, including the following:

- **Soil treatment** is important because damage to soil structure and function is a common and serious problem in degraded semi-arid and arid areas. Arid soil, compacted soil, and nutrient-poor soil may need to be improved by adding organic amendments, such as leaf and litter compost, composted manure, biosolids, or mulch that is certified contaminant and weed-free. These amendments could help bind recalcitrant organic compounds and metals and increase the much-needed water holding capacity and fertility. Other measures to improve soil structure and function include soil surface treatments, such as creating pits in soil, to improve water retention in arid land and imprinting, to increase soil moisture and gully control to improve plant establishment.
- **Water availability** for plants may improve if the ground is shaped to collect and retain water. Transplanted seedlings may need limited irrigation to survive until established. Species selections can also be adapted to local hydrology. Too much irrigation may encourage invasive weeds, leave salts at the soil surface that kill plants, or cause infiltration into subsurface contaminated materials.
- **Seed selection** for arid areas is hampered by the limited availability of commercial stocks of dry land seeds. If possible, the project manager may hire a commercial seed collector to collect seed from the local area or an area with similar climate. The alternate collection area needs to be within a 100-mile radius and 500 feet of the altitude of the area to be planted; where the average rainfall is within two inches per year of the annual rainfall for the area; and have similar soil characteristics (Department of the Interior [DOI] 1995). Seed testing can help cleanup project managers ensure that the seeds are of high quality. Proper seed storage will also help maintain the seed's viability until sowing. Visit the Plant Conservation Alliance Web site for a directory of restoration experts and native seed suppliers (www.nps.gov/plants
- **Planting techniques** primarily include direct seeding and transplanting. Direct seeding is generally less expensive. However, in dry areas this technique is more vulnerable to seed loss from exposure to wind, insects, and rodents, as well as declines in germination rates and plant growth because of insufficient rainfall in the months following planting. The installation of an erosion blanket consisting of straw or coco fiber with biodegradable netting can help prevent seed loss and retain moisture while plants are established. Cleanup project managers may also consider using collected seed to grow container plants for drier areas. If container plants are used, additional time will be necessary to allow the plants to germinate and achieve the desired growth in a greenhouse or nursery before planting. Using container plants can be costly and labor intensive. Because plant losses usually occur, it is prudent to budget for monitoring and replacement.

7.0. Long-Term Stewardship Considerations

Cleanups are risk-based and, when waste is left in place, long-term stewardship is necessary to ensure protectiveness of the remedy; therefore, long-term stewardship responsibilities are an integral part of the cleanup process. O&M activities through responsible stewardship protect the integrity of the cleanup and the functioning of the associated ecosystems after cleanup completion. For example, at the Woodlawn Landfill Superfund Site, WHC and Bridgestone Americas Holding, Inc. conducted ecological revitalization activities at the site to create wildlife habitat. Local volunteers manage the site. In addition, Chicago's pocket park project highlighted earlier in Section 2 incorporated (1) ICs and (2) community involvement in site planning and maintenance, which reduced costs and helped ensure the success of ecological revitalization. See Appendix A for case studies regarding these sites.

There are four major components for a successful O&M program:

- Plan early for long-term stewardship
- Identify and complement general O&M activities
- Establish a monitoring program
- Use ICs

Long-Term Stewardship. EPA's co-regulatory partners, including states, local governments, and tribes, have increasing responsibility and oversight for property assessment and cleanup planning. This property knowledge is particularly important for long-term stewardship as state voluntary cleanup programs and property owners typically have primary responsibility for carrying out maintenance of engineering controls and ICs for the long-term. Therefore, it is essential to prepare for safeguarding the effectiveness of the ecological revitalization activities as early in the cleanup planning process as possible. Regardless of who is responsible for O&M, stakeholders can make agreements to have general maintenance tasks as well as those specific to ecological revitalization implemented by property owners, a local government agency, Trustees, or the community. It may be practical to have the same organization undertake general O&M activities as well as those relating specifically to the ecosystem. For example, at the Silver Bow Creek/Warm Springs Ponds Superfund Site in Montana, the Montana Department of Fish, Wildlife, and Parks, a Trustee, conducts many general and specific monitoring and maintenance tasks (see case study in Appendix A).

Stakeholder Collaboration at a Former Refinery in Casper, Wyoming

Stakeholders are successfully achieving cleanup of a BP former refinery in Casper, Wyoming through a collaborative process. The group redeveloped the former refinery into a business park and golf course where the wetland treatment system also functions as a golf course water hazard. To reach agreement on the cleanup, BP worked closely with stakeholders, including the local Audubon Society and the community. The Audubon Society used its local expertise to help determine an appropriate shoreline elevation to maintain the wetlands and mud flats. See Appendix A for a case study regarding this site.

Cleanup project managers can also enlist a local group or guardian to conduct long-term stewardship of a property. Such groups are committed to follow-through and have knowledge of local conditions. They can also monitor the ecological revitalization component and look for early signs of any emerging issues. Local government agencies can also provide expertise, equipment, supplies, or other resources to help the local community or group conduct long-term stewardship; this can reduce costs, provide interpretive educational benefits, and help encourage a sense of property ownership by the community.

General O&M Activities. In some cases, appropriately designed ecosystem revitalization may be self- sustaining and need little or no maintenance after an initial establishment period. In most cases, however, O&M will be necessary. O&M activities depend on the type of cleanup as well as the ecological revitalization component and, depending on the situation, are often necessary for a long period of time (up to 20, 50, or 100 years). O&M for the overall cleanup typically includes inspection, sampling and analysis, routine maintenance and small repairs, and reporting, as necessary. Cleanup project managers can incorporate ecological revitalization measures into each of these tasks.

- **Inspection needs to occur on a regular basis.** Inspectors can also perform non-routine inspections after unusual events such as earthquakes or large storms. Typically, inspectors check for invasive species, erosion, and dead or dying vegetation, among other items, when assessing the ecological revitalization component of the cleanup. For properties with cover systems in place, inspectors also check for settling, burrowing animals, and pooling water. Cleanup project managers typically include performance standards to measure the success of the project, as well as a detailed description of how team members will conduct inspections, sampling, and maintenance activities.
- **Regular sampling and analysis** helps monitor habitat, ground water, and surface water quality. Monitoring habitat indicators such as plant species composition and percentage of cover helps to determine the success of the revitalization measures. In addition, making a determination of the amount of invasive plant species in the area helps to ensure that they are not overtaking the area. Sampling and analysis includes collecting and chemically analyzing water samples from surface water, wetlands, or ground water wells; soil samples may also be collected and analyzed to evaluate soil conditions. For properties with cover systems in place, sampling would include leachate formation and gas release concentrations. The frequency of sample collection can vary widely and needs to be determined on a property-specific basis.
- **Routine maintenance** may consist of simple activities such as burning, using herbicide, or mowing to control invasive species; maintaining a cover; or repairing perimeter fencing. On properties that have operating treatment plants, routine maintenance may be more complex and may need a full- or part-time plant operator. Typical activities include operating ground water and gas treatment systems, repairing erosion damage, and maintaining rainwater collection and diversion systems. Based on inspection results and plant species composition and cover at the revitalization area, reseeding or replanting may be necessary as well as periodic mowing or controlled burns. Manual or natural controls or herbicides or insecticides applications can also control invasive plants and undesirable insects and diseases. For additional information on maintaining a variety of habitat types, review ITRC's Planning and Promoting Ecological Land Reuse of Remediated Sites (ITRC 2006).
- **Reporting** requirements depend on the cleanup program, and cleanup project managers generally write and submit reports to regulatory authorities after both routine and non-routine inspections. The reports typically include information on the general condition of the cleanup measures, test results from samples collected, and

operational data from treatment processes (for example, ground water extraction rate, gas flow rate).

Loring Air Force Base in Maine

Cleanup project managers for Loring Air Force Base consulted with the U.S. Fish and Wildlife Service (USFWS) to identify useful indicator species such as dragon fly nymphs, midge flies, dace minnow, and brook trout to monitor the recovery of the stream system after remedial activities. These species were selected because they are sensitive to contaminants and are quick to manifest symptoms of exposure. See Appendix A for additional case study information.

Monitoring Program. A monitoring program, established as part of post-cleanup activities, evaluates the effectiveness of the cleanup in restoring ecological function and reducing ecological risks (EPA 1998, 1999a). Information from baseline surveys and ERAs conducted during the planning process can be the starting point for developing the monitoring program. For example, periodic monitoring of sediment contamination and benthic communities following the removal of contaminated sediment in a stream can provide indications of the protectiveness of the cleanup features as well as the ecosystem's recovery to a more natural condition. At the Revere Chemical Company Superfund Site in Pennsylvania, ground water and stream monitoring is used to evaluate the risks of heavy metals getting into the ground water and migrating off site. Cleanup project managers also use the monitoring program to help evaluate the recovery of important aquatic species. Monitoring habitat indicators such as plant species composition and percent cover could indicate the success of the revitalization measures. See Appendix A for a case study regarding this site.

Designing and Implementing Institutional Controls

Many factors may influence the design and implementation of ICs, such as state policies, whether the property is a federal facility, or whether regulatory authorities, such as RCRA or CERCLA, are involved. An EPA guide addresses many of these issues (EPA 2000). Visit the following Web site to view the guide: http://epa.gov/superfund/policy/ic/guide/guide.pdf

Institutional Controls. ICs are designed to limit land or resource use, and provide information to help modify or guide human behavior, and complement engineering controls. They can also protect ecological revitalization properties by restricting public access to parts of a property that are particularly sensitive to erosion or contain sensitive or establishing habitats; or to achieve human protectiveness or other revitalization goals. A key to success is to identify and evaluate as much information as possible about the needed ICs early in the planning process. Generally, major considerations with IC use at ecological revitalization properties include the following:

- **Consider what the IC is intended to accomplish and establish clear objectives.** A common IC objective for ecological purposes involves controlling human activities in a particular area that could potentially interfere with sensitive habitats or the ecosystem balance that supports the cleanup features.

- **Consider the appropriate types of ICs.** These can include governmental controls (zoning, building codes, and ground water use restrictions), proprietary controls (easements, covenants, and conservation trusts), enforcement tools (consent decrees and administrative orders), and informational devices (fishing advisories, deed notices, and state registries of contaminated properties). For example, a conservation easement for catch and release fishing and a local health department fishing advisory could accomplish the same IC objective to reduce fish consumption. For information about different types of ICs, see EPA's guide titled Institutional Controls: A Site Manager's Guide to Identifying, Evaluating, and Selecting Institutional Controls at Superfund and RCRA Corrective Action Cleanups at *http://epa.gov/superfund/policy/ic/guide/guide.pdf* (EPA 2000).
- **Ensure that the specified ICs are effective and remain in place over the long term** through proper implementation, monitoring, and enforcement. For example, at the Silver Bow Creek Superfund Site in Butte, Montana, the Montana Department of Fish, Wildlife, and Parks enforces a fish consumption prohibition. In addition, at the BP Former Refinery in Casper, Wyoming, project managers implemented several ICs including a "use control area" through a resolution to limit use on the property, a ground water restriction area, and a soil management overlay district. Within one of these defined areas, a constructing entity has to contact the state or BP if they have been issued a building permit. See Appendix A for additional information on these case studies.

REFERENCES

[1] Calumet Ecotoxicology Roundtable Technical Team. (2007). *Calumet Area Ecotoxicology Protocol, Protecting Calumet's Plants and Animals*. June.

[2] Interstate Technology and Regulatory Council (ITRC). (2006). *Planning and Promoting Ecological Land Reuse of Remediated Land ECO-2*. Washington D.C.: ITRC and Wildlife Habitat Council (WHC). www.itrcweb.org/Documents/ECO-2.pdf.

[3] Kent, D. M. (ed.). (1994). *Applied Wetlands Science and Technology*. Boca Raton, FL: CRC Press.

[4] National Research Council (NRC). (1992). *Restoration of Aquatic Ecosystems: Science, Technology, and Public Policy*, Washington, D.C.: National Academy Press.

[5] Northeast-Midwest Institute and the National Brownfields Coalition. (2007). *EPA Brownfields Program— Issues and Opportunities. Petroleum/UST Brownfield Cleanups. September. www.nemw.org/ petroleum%20issue%20opportunity%20brief.pdf.*

[6] U. S. Department of Interior. (1995). A Beginner's Guide to Desert Restoration, National Park Service, *Desert Restoration Task Force*, Denver Service Center.

[7] U.S. Environmental Protection Agency (EPA). (1988). *America's Wetlands: Our Vital Link between Land and Water*. OPA-87-016.

[8] EPA. (1991). *Design and Construction of RCRA/CERCLA Final* Covers. EPA 625-4-

91-025.

[9] EPA. (1993). *Environmental Fact Sheet: Controlling the Impacts of Remediation Activities in or Around Wetlands*. Office of Solid Waste and Emergency Response. EPA 530-F-93-020.

[10] EPA. (1994). *Considering Wetlands at CERCLA Sites. Office of Solid Waste and Emergency Response*. EPA 540-R-94-019. www.epa.gov/ superfund/policy/remedy/ pdfs/540r-94019-s.pdf.

[11] EPA. (1995). *Ecological Restoration: A Tool to Manage Stream Quality*. November. EPA 841-F-95-007. www.epa.gov/OWOW/NPS/Ecology (viewed on 01/12/2009).

[12] EPA. (1998). *Guidelines for Ecological Risk Assessment Guidance*. EPA 630-R-95-002F. April. http://cfpub.epa.gov/ncea/cfm/ recordisplay.cfm?deid=12460 (viewed on 01/12/2009).

[13] EPA. (1999a). *Issuance of Final Guidance: Ecological Risk Assessment and Risk Management Principles for Superfund Sites*. OSWER 9285.7-28. October. www.epa.gov/oswer/riskassessment/ ecorisk/pdf/final99. pdf.

[14] EPA. (1999b). *RCRA Cleanup Reforms. Faster, Focused, More Flexible Cleanups*. OSWER. EPA 530-F99-018. July. www.epa.gov/epaoswer/ hazwaste/ca/reforms

[15] EPA. (2000). *Institutional Controls: A Site Manager's Guide to Identifying, Evaluating, and Selecting Institutional Controls at Superfund and RCRA Corrective Action Cleanups*. EPA 540-F-00-005. http://epa.gov/superfund/policy/ic/guide/guide.pdf.

[16] EPA. (2001). *RCRA Cleanup Reforms, Reforms II: Fostering Creative Solutions*. OSWER. EPA 530-F-01- 001. January.

[17] EPA. (2002a). *Reusing Cleaned Up Superfund Sites: Commercial Use Where Waste is Left on Site*. EPA540-K-01-008. February. www.epa.gov/superfund/programs/recycle/ pdf/c_reuse.pdf.

[18] EPA. (2002b). *Region 5 Corrective Action Sites: Majority of RCRA Federal Lead Sites Have Reuse Potential*. Waste, Pesticides and Toxics Division. February. www.epa.gov/swerosps/rcrabf/pdf/surveyfs.pdf.

[19] EPA. (2005). *Ecological Reuse of Remediated Sites: Some Resources*. Office of Solid Waste. December. www.clu-in.org/conf/tio/ ecoresources3/prez/ecoresources3pdf.pdf.

[20] EPA. (2006a). *2006–2011 EPA Strategic Plan—Charting Our Course*. www.epa.gov/ ocfo/plan/2006/entire_report.pdf.

[21] EPA. (2006b). *Measuring Revitalization of Contaminated Properties in America's Communities: Past Accomplishments and Future Opportunities*. OSWER. EPA 500-R-06-002. September.

[22] EPA. (2006c). *Frequently Asked Questions about Ecological Revitalization of Superfund Sites*. EPA 542-F06-002. www.clu-in.org/ download/remed/542f06002.pdf.

[23] EPA. (2006d). *Revegetating Landfills and Waste Containment Areas Fact Sheet*. EPA 542-F-06-001. www.clu-in.org/download/remed/ revegetating_fact_sheet.pdf.

[24] EPA. (2006e). *Interim Guidance for OSWER Cross-Program Revitalization Measures*. October.
www.epa.gov/landrevitalization/docs/cprmguidance-10-20-06covermemo.pdf.

[25] EPA. (2007a). *BP Former Refinery*, Casper, Wyoming. October. www.epa.gov/waste

[26] EPA. (2007b). *DuPont Remington Arms Facility*. October. *http://epa.gov/epawaste/ hazard/correctiveaction/pdfs/rem11-07.pdf.*

[27] EPA. (2007c). *Ecological Revitalization and Attractive Nuisance Issues*. EPA 542-F-

06-003. www.cluin.org/download/remed/542f06003.pdf.

[28] EPA. (2007d). *The Use of Soil Amendments for Remediation, Revitalization, and Reuse.* EPA 542-R-07- 013. December. www.clu-in.org/download/remed/epa-542-r-07-013.pdf.

[29] EPA. (2007e). *Memorandum Regarding the Underground Storage Tanks Program Plan for Implementing OSWER's Cross-Program Revitalization Measures.* February.

[30] EPA. (2007f). *Guidance for Documenting and Reporting RCRA Subtitle C Corrective Action Land Revitalization Indicators and Performance Measures.* February 21. www.epa.gov/reg3wcmd/ca/pdf/finalRCRA CPRM guidance2 21 07.pdf

[31] EPA. (2007g). Memorandum Regarding the Update and Next Steps for Superfund and Federal Facilities Cross-Program Revitalization Measures (CPRM) Implementations. From James Woolf ord, Director, OSRTI, and John Reeder, Director, FFRRO, to Superfund National Policy Managers, Regions 1–10. June 20.

[32] EPA. (2008a). *Superfund Redevelopment.* www.epa.gov/superfund/ programs/recycle/ info/index.html(viewed on 01/12/2009).

[33] EPA. (2008b). *Land Revitalization—Integrating Reuse Into Cleanup.* www.epa.gov/oswer/landrevitalization/aiintegratingreuse1.htm(viewed on 01/12/2009).

[34] EPA. (2008c). *Underground Storage Tanks.* www.epa.gov/swerust1 (viewed on 01/12/2009).

[35] EPA. (2008d). *U.S. EPA's Petroleum Brownfields Action Plan: Promoting Revitalization and Sustainability. October. www.epa.gov/oust/rags/petrobfactionplan. pdf.*

[36] EPA. (2008e). *RCRA Brownfields Prevention Initiative.* www.epa.gov/ swerosps/rcrabf (viewed on 01/12/2009).

[37] EPA. (2008f). OUST. *Cleaning Up and Reusing Abandoned Gas Station Sites.* www.epa.gov/oust/rags/index.htm (viewed on 01/12/2009).

[38] EPA. (2008g). *River Corridor and Wetlands Restoration.* Benefits of Restoration. www.epa.gov/owow/wetlands (viewed on 01/12/2009).

[39] EPA. (2008h. *Atlas Tack Corporation Superfund Site*, Fairhaven, Massachusetts. Final Interim Remedial Action Report (O&F Completion Report) for Phases II and III. September. www.epa.gov/region1/ superfund/sites/atlas/ATLAS TACK TARGET SHEET.pdf

[40] EPA. (2008i). Information based on monthly database reports generated in February 2008 and communicated via e-mail by Stacy Swartwood, OBLR to Tetra Tech EM Inc., dated September 26, 2008.

[41] EPA. (2008j). Green Remediation: Incorporating Sustainable Environmental Practices into Remediation of Contaminates Sites. EPA 542-R-08-002. April. www.clu-in.org/download/remed/GreenRemediation-Primer.pdf

[42] EPA. (2009). Office of Solid Waste and Emergency Response FY2009 National Program Manager's Guidance (draft). February. www.epa.gov/ocfo/npmguidance/ oswer/2009/draft npm guidance.pdf

[43] U.S. Fish and Wildlife Service (USFWS). (1984). *An Overview of Major Wetlands Functions and Values.* FWS OBS-84-18.

[44] Wildlife Habitat Council (WHC). (2008). *Brownfields/Remediation. www.wildlifehc. org/brownfields/index.cfm* (viewed on 01/12/2009).

APPENDIX A. ECOLOGICAL REVITALIZATION CASE STUDIES

Property Name and Location	Property Type	Cleanup Type	Revitalization/Reuse Component	Problems/Issues	Solutions	Point of Contact	Notes/Links*
REGION 1							
Atlas Tack Superfund Site, Fairhaven, MA	Superfund Manufacturing Facility	Ground water contaminated with cyanide and toluene that leached from the site lagoon and soils contaminated with VOCs, heavy metals, pesticides, PCBs, and PAHs were cleaned up by removing buildings, contaminated soil, and sediment.	The cleanup preserved as much of the wetland sediment as possible and provided the necessary mix of fresh and salt water sources to create a functioning wetland, in addition to protecting human health and the environment.	1. The original ROD contained sediment cleanup values that would require complete excavation of the entire marsh. 2. The initial remediation plan included lowering the ground water table to prevent it from flowing through residual contamination.	1. The bioavailability study showed that it was not necessary to remove all sediments, and therefore only necessary sediment was removed, thereby preserving the marsh to the extent possible. 2. The remediation approach was re-evaluated during wetland design, and risks from ground water flowing beneath the site were minimal.	Elaine Stanley, RPM EPA Region 1 1 Congress Street Suite 1100 Mail Code: HBO Boston, MA 02114-2023 617-918-1332 stanley.elainet@epa.gov	http://www.epa.gov/ ne /superfund/sites/atlas/

(Continued)

Property Name and Location	Property Type	Cleanup Type	Revitalization/Reuse Component	Problems/Issues	Solutions	Point of Contact	Notes/Links*
Fort Devens: OU2 Devens Consolidation Landfill, Sudbury, MA	Superfund Military Base	Numerous small historical landfills were remediated and the waste was consolidated in a new state-of-the-art landfill. Soils and debris disposed at the Devens Consolidation Landfill included those contaminated with petroleum, pesticides, PCBs, PAHs, and asbestos. A total of approximately 365,000 cubic yards of	Three of the historic landfills had waste or debris in wetland areas. For these areas, the remedy included waste and debris removal, followed by wetland restoration. The wetlands were restored by backfilling with clean fill and manufactured wetland soil. Materials were stabilized with a custom wetland seed mix, in accordance with a Habitat Restoration Work Plan. The site was monitored and evaluated during the next three growing seasons to ensure it achieved restoration success measures.	Not specified	Not specified	Ginny Lombardo, RPM EPA Region 1 1 Congress Street Suite 1100 Mail Code: HBT Boston, MA 02114-2023 617-918-1754 lombardo.ginny@epa.gov	http://yosemite. epa. gov/r1/npl pad.nsf/51dc4f173 cecf51d85 256adf004c7cc8/ df7d 91 0ff9a9 3fab8525691f006 3f6c 9!Open Document&Highlight=0, devens

(Continued)

Property Name and Location	Property Type	Cleanup Type	Revitalization/Reuse Component	Problems/Issues	Solutions	Point of Contact	Notes/Links*
		waste was disposed of in the new landfill.The historic landfill sites were then backfilled and regraded to restore the sites to pre-construction conditions.					
Fort Devens: OU9 AOC 57, Sudbury, MA	Superfund Military Base	AOC 57 consists of 2 areas that were affected by stormwater runoff and wastes from vehicle maintenance activities at a historic storage yard upgradient of the site. The areas are	Soil excavation at one of the areas included excavation within delineated wetland areas along Cold Spring Brook. The remedy required that the wetland areas be restored in accordance with an appropriate mitigation and restoration plan and that the wetland restoration area be .	Not specified	Not specified	Ginny Lombardo, RPM EPA Region 1 1 Congress Street Suite 1100 Mail Code: HBT Boston, MA 02114-2023 617-918-1754 lombardo.ginny@epa.gov	http://yosemite epa .gov/r1/npl pad.nsf/51dc4f 173 ceef 51d85 256adf004c7cc8/ df 7d9 10ff9a9 3fab8525691f006 3f6 c9! Open Document&Hig-hlight =0, deven S

(Continued)

Property Name and Location	Property Type	Cleanup Type	Revitalization/Reuse Component	Problems/Issues	Solutions	Point of Contact	Notes/Links*
		sloped along Cold Spring Brook. Soils and ground water were contaminated with petroleum hydrocarbons, chlorinated VOCs, PCBs, and arsenic. Contaminated soils were removed and disposed off-site, and ground water will be remediated via MNA.	monitored for 5 years to ensure that restoration success measures were achieved.				
GE-Housatonic River, Pittsfield, MA	Superfund Manufacturing Facilities	Site remediation involved clean up of Housatonic River sediments and floodplain	GE is providing economic aid to the City of Pittsfield for 10 years and making upgrades to the Housatonic River, its floodplain, and Silver	Issues relating to flood storage compensation are under discussion with EPA.	Not specified	Thomas Hickey, Jr. Pittsfield Economic Development Authority 81 Kellogg Street Pittsfield, MA 01201 413-494-7332 thickey@peda.cc	http://www.epa.gov/region1/ge/redevelopment.html

(Continued)

Property Name and Location	Property Type	Cleanup Type	Revitalization/Reuse Component	Problems/Issues	Solutions	Point of Contact	Notes/Links*
		soils contaminated with PCBs and other hazardous substances. Remediation included excavating and disposing of sediment and soil and full-scale capping of Silver Lake.	Lake that will have aesthetic value and enhance local habitat.				
Industri-Plex Site, Woburn, MA	Superfund Manufacturing Facility	The remedy included remediating approxi-mately 110 acres of soil contaminated with lead, arsenic, and chromium; demolishing onsite buildings; and constructing	Wetlands and open space were created adjacent to redeveloped areas, which included a regional transportation center, highway interchange, and land developed for retail and commercial use.	None	None	Joseph LeMay, RPM EPA Region 1 1 Congress Street Suite 1100 Mail Code: HBO Boston, MA 02114-2023 617-918-1323 lemay.joe@epa.gov	http://yosemite. epa. gov/r1/npl pad.nsf/f52fa5c 31fa8 f5c8852 56adc0050b631 /1E 8F7 D6FFC D9B61B85256 A0F 00067136?OpenDocument

(Continued)

Property Name and Location	Property Type	Cleanup Type	Revitalization/Reuse Component	Problems/Issues	Solutions	Point of Contact	Notes/Links*
		clay, soil, and synthetic layers, concrete foundations, and asphalt to cover contamination. In addition, gases at a hide pile were collected and treated, and wetlands and open spaces were created.					
Iron Horse Park, North Billerica, MA	Superfund Manufacturing Facility Landfill	On-site ground water and surface water were contaminated with organic and inorganic chemicals, asbestos, and heavy metals. The soil at the site was	Wetlands were restored.	Not specified	Not specified	Don McElroy EPA Region 1 1 Congress Street, Suite 1100 Mail Code: HBO Boston, MA 02114-2023 617-918-1326 mcelroy.don@epa.gov	http://yosemite.epa. gov/r1/npl pad.nsf/51dc4f 173cec f51d85 256adf004c7ec8/c 334 fff032ec c1c78525691f00 63f 6d0 ?Open Document

(Continued)

Property Name and Location	Property Type	Cleanup Type	Revitalization/Reuse Component	Problems/Issues	Solutions	Point of Contact	Notes/Links*
		contaminated with PCBs, petro-chemi-cals, and heavy metals. Remediation activities included capping on-site landfills and exca-vating and removing contaminated soil and sediment.					
Jamaica Island Landfill OU3, Kittery, ME	Superfund Remedial Action Landfill	A variety of organic and inorganic constituents were detected in soil and ground water and included VOCs, SVOCs, PCBs, pesticides, metals, and	Wetlands were constructed.	Minimizing the effect on existing mudflats in the area and locating appropriate backfill to	Not specified	Fred Evans, RPM Navy Portsmount Naval Shipyard Kittery, ME 03904 610-595-0567 ext. 159 evansfj@efanc.navfac.navy.mil	http://www.wild-lifehc. org/eweb editpro/items/O57 F3078.pdf

(Continued)

Property Name and Location	Property Type	Cleanup Type	Revitalization/Reuse Component	Problems/Issues	Solutions	Point of Contact	Notes/Links*
		petroleum hydrocarbons. Remediation included installation of a cap and shoreline erosion controls.		maximize the potential for success.			
Loring Air Force Base, Northeastern ME	Superfund Air Force Base	Ground water contaminated with VOCs and fuel-related compounds and surface water and sediment contaminated with VOCs, PCBs, and heavy metals were remediated. Activities included capping on-site landfills and excavating and removing	Boulders and cobbles from the streambed and nearby trees larger than 15 centimeters in diameter that were removed during cleanup were later used in stream reconstruction, after completion of cleanup activities. Reuse of native materials significantly reduced the cost of restoration materials.	Not specified	Not specified	Mike Daly, RPM EPA Region 1 1 Congress Street Suite 1100 Mail Code: HBT Boston, MA 02114-2023 617-918-1386 daly.mike@cpa.gov	http://cfpub.cpa.gov/supercpad/cursites/csitinfo.cfm? id=0101074

(Continued)

Property Name and Location	Property Type	Cleanup Type	Revitalization/Reuse Component	Problems/Issues	Solutions	Point of Contact	Notes/Links*
		contaminated soil and sediment.					
Materials Technology Laboratory, Watertown, MA	Superfund Arsenal	Remediation included removal and off-site disposal of contamination sources related to weapons and ammunition manufacture and storage, and demolition and cleanup of the nuclear reactor, including radiological contamination, PAHs, PCBs, and pesticides.	Wetlands restoration was completed adjacent to the redeveloped area. Fifty-five acres of the property have been used to build the Arsenal Mall, Harvard Community Health Center, Arsenal Apartments, a public park with walking and bike trails, and a playground.	Not specified	Not specified	Christine Williams, RPM EPA Region 1 1 Congress Street Suite 1100 Mail Code: HBT Boston, MA 02114-2023 617-918-1384 williams.christine@epa.gov	http://yosemite.epa.gov/r1/npl pad.nsf/701b6886f1 89ccae85256bd20014c93d/d98829ad20c19d6f852568ff005adb08!OpenDocument

(Continued)

Property Name and Location	Property Type	Cleanup Type	Revitalization/Reuse Component	Problems/Issues	Solutions	Point of Contact	Notes/Links*
Pease Air Force Base, Portsmouth, NH	Superfund Air Force Base	Soils and ground water were contaminated with solvents and fuel.	A wildlife refuge was created in addition to a public airport.	Not specified	Not specified	Mike Daly, RPM EPA Region 1 1 Congress Street Suite 1100 Mail Code: HBT Boston, MA 02114-2023 617-918-1386 daly.mike@epa.gov	http://yosemite.epa.gov/r1/nplpad.nsf/f52fa5c31fa8f5c885256adc0050b631/9E95FBAD0CEC73E0852568FF005ADB09?OpenDocument
Saco Municipal Landfill, Saco, ME	Superfund Landfill	Soil and ground water contaminated from landfill activities were remediated.	A portion of the site adjacent to the redeveloped area was reserved for a wetland. The site is ready for reuse and the City of Saco plans to develop a community recreation area for hiking, biking, ice skating, and soccer.	Not specified	Not specified	Ed Hathaway, RPM EPA Region 1 1 Congress Street Suite 1100 Boston, MA 02114-2023 617-918-1372 hathaway.ed@epa.gov	http://cfpub.epa.gov/supercpad/cursites/csitinfo.cfm?id=0101010
Tibbetts Road Site, Barrington, NH	Superfund Rural/Farmland	Site soils and ground water were contaminated by chlorinated and non-chlorinated solvents.	The wooded phytoremediation area is providing increased biodiversity through new wildlife habitat for various birds and small mammals.	Not specified	Not specified	Jerome S. Amber, P.E. Ford Motor Company, retired 248-765-1044 jamber@comcast.net	http://www.wildlifehc.org/cwebeditpro/items/O57F3072.pdf

(Continued)

Property Name and Location	Property Type	Cleanup Type	Revitalization/Reuse Component	Problems/Issues	Solutions	Point of Contact	Notes/Links*
		Remediation included source removal, building demolition, water supply extension, and phytoremediation.					
REGION 2							
Asbestos Dump, Millington, NJ	Superfund Landfill	Asbestos from 4 sites was collected, consolidated, and treated on-site to prevent release of contaminants. A soil cover was then placed over the site.	A barn was converted into an environmental awareness center. Most of the property will be preserved and will help expand the Great Swamp National Wildlife Refuge.	Not specified	Not specified	Carla Struble, RPM EPA Region 2 290 Broadway New York, NY 10007-1866 212-637-4322 struble.carla@epa.gov	http://yosemite.epa.gov/opa/admpress.nsf/b853 d6fc004accbf852572a000656840/3f082ac6d59bb9ac85257165006bc507!OpenDocument

(Continued)

Property Name and Location	Property Type	Cleanup Type	Revitalization/Reuse Component	Problems/Issues	Solutions	Point of Contact	Notes/Links*
DeRewal Chemical Co., Kingwood Township, NJ	Superfund Chemical Company	Contaminated soil and ground water from chemical spills was cleaned up through excavation and treatment of soil and extraction and treatment of ground water.	The site now contains walking, canoe, and biking trails, and bird watching opportunities. The Kingwood Township also plans to convert a house on the site into a historical, environmental, and recreational center.	Not specified	Not specified	EPA Region 2 290 Broadway New York, NY 10007-1866	http://www.epa gov/region02/s uperfund/npl /0200792c.pdf
Lipari Landfill, Pitman, NJ	Superfund Landfill	A slurry wall and cap were constructed for the landfill, which accepted wastes contaminated with VOCs and heavy metals. A ground water and leachate P&T system	Revitalization included recreational use of a park and lake as well as development of streams and marshes.	In the ROD for OU2, changes in the remedy flow rates, equipment sizes, and estimated costs in design were made to the on-site containment facilities. The ROD for OU3 included changes to the soil and sediment volumes handled	Changes in the ROD did not change the functionality of the remedies.	Melissa Friedland EPA HQ Ariel Rios Building 1200 Pennsylvania Avenue Mail Code: 5204P Washington, DC 20460 703-603-8864 friedland.melissa@epa.gov	http://cfpub .epa. gov/supercpa d/cursites/ csitinfo. cfm?id=020 0557

(Continued)

Property Name and Location	Property Type	Cleanup Type	Revitalization/Reuse Component	Problems/Issues	Solutions	Point of Contact	Notes/Links*
		was installed,and contaminated soil and sediment were excavated and treated.		and methods for removing sediment.			
Marathon Battery, Cold Spring, NY	Superfund Manufacturing Facilities	The factory and surrounding soils, a nearby marsh, and adjacent river sediments were contaminated with heavy metals. Remediation included excavating, capping, and restoring the marsh; excavating contaminated soils;	The marsh is now used for recreational and educational purposes, and the factory grounds are ready for redevelopment.	Difficulties included experienced goose predation, destructive ice flows, invasive plant species, and bare areas due to differential settlement within the marsh.	Each problem was dealt with individually. Some areas were replanted, coir logs were used to encourage natural plant coverage and sediment build-up in bare areas, and beetles were used to retard the growth of invasive species.	Pam Tames, RPM EPA Region 2 290 Broadway New York, NY 10007-1866 212-637-4255 tames.pam@epa.gov	http://www .epa .gov/Region 2/s uperfund/npl/ 0201491c.pdf

(Continued)

Property Name and Location	Property Type	Cleanup Type	Revitalization/Reuse Component	Problems/Issues	Solutions	Point of Contact	Notes/Links*
		dredging cove and river sediments; and demolishing the plant.					
Myers Property Superfund Site, Hunterdon County, NJ	Superfund Manufacturing Facility	Soil and ground water contaminated with VOCs, pesticides, semiVOCs, metals, and dioxins were cleaned up by excavating contaminated soil and sediment, treating soil, and extracting and treating ground water.	RPMs are saving existing trees above a certain size in areas with low levels of contamination by hand digging around the roots to a depth of six inches. Excavated soil will be replaced with clean topsoil from off site.	Subsurface soil contamination remains, so if a tree falls, contaminated soil could be exposed.	The property will be monitored in case large trees fall and expose soils deeper than six inches.	Stephanie Vaughn, RPM EPA Region 2 290 Broadway, 19th Floor New York, NY 10007-1866 212-637-3914 vaughn.stephanie@epa.gov	http://www.epa .gov/region02/s uperfund/npl/ 0200774c.pdf
REGION 3							
Army Creek Landfill, DE	Superfund Landfill	Remediation of soil and ground water contaminated with VOCs,.	Native vegetation was planted to create a bird and wildlife habitat. In addition, discharge	Not specified	Not specified	Deb Rossi, RPM EPA Region 3 1650 Arch Street Mail Code: 3HS23 Philadelphia, PA 19103-2029 215-814-3228	http://www.epa. gov/superfund/ programs/ recycle /live/casestu

(Continued)

Property Name and Location	Property Type	Cleanup Type	Revitalization/Reuse Component	Problems/Issues	Solutions	Point of Contact	Notes/Links*
		chromium, and mercury included a multi-layer protective cover over a municipal and industrial landfill and a ground water treatment system. Army Creek was also contaminated with cadmium, chromium, mercury, iron, and zinc	pipes from the ground water treatment system were routed to create wetlands to help prevent flooding and create additional habitat.			rossi.debra@epa.gov	
Avtex Fibers, Front Royal, VA	Superfund Manufacturing Facilities	The principle contaminants found in the ground water were carbon disulfide, ammonia, arsenic, antimony, phenol, and high pH. Arsenic, lead,	The site was used to create a river conservancy park, active recreation park, and an eco-business park.	Not specified	Not specified	Bonnie Gross, RPM EPA Region 3 1650 Arch Street Mail Code: 3HS23 Philadelphia, PA 19103-2029 215-814-3229 gross.bonnie@epa.gov	dyarmycreek.html accomp/success/avtex.htm

(Continued)

Property Name and Location	Property Type	Cleanup Type	Revitalization/Reuse Component	Problems/Issues	Solutions	Point of Contact	Notes/Links*
		and PCBs have been identified in soils. PCBs associated with the plant were also detected in the Shenandoah River. Remediation was completed by demolishing or decontaminating onsite buildings, removing and treating onsite hazardous and nonhazardous chemical waste, excavating contaminated soil and debris, and constructing a low-flow wastewater treatment system.					

(Continued)

Property Name and Location	Property Type	Cleanup Type	Revitalization/Reuse Component	Problems/Issues	Solutions	Point of Contact	Notes/Links*
Berks Landfill, Berks County, PA	Superfund Landfill	Ground water was contaminated with VOCs and metals. The remedy included ICs, long-term monitoring of ground water, operation and maintenance of the leachate system, and repair to the landfill cap.	The former residential property at the site is being reused as open green space with trees and vegetation. ICs were implemented in order to prevent on-site ground water use and to protect the landfill cap.	Not specified.	Not specified	Kristine Matzko EPA Region 3 1650 Arch Street Mail Code: 3HS21 Philadelphia, PA 19103-2029 215-814-5719 matzko.kristine@epa.gov	http://www.epa.gov/superfund/sites/fiveyear/f05-03018.pdf
Butz Landfill, Monroe County, PA	Superfund Landfill	A former municipal dump contaminated the ground water with a solvent, TCE, and other organic compounds. Nearly 82,720,000 gallons of water	Revitalization involved creating wetlands to mitigate potential loss of wetlands caused by the P&T system.	Not specified	Not specified	Romuald A. Roman, RPM EPA Region 3 1650 Arch Street Mail Code: 3HS22 Philadelphia, PA 19103-2029 215-814-3212 roman.romuald@epa.gov	*http://www.epa.gov/reg3hscd/s* uper/sites/PAD 981034705/

(Continued)

Property Name and Location	Property Type	Cleanup Type	Revitalization/Reuse Component	Problems/Issues	Solutions	Point of Contact	Notes/Links*
		were treated using a P&T system.					
Chisman Creek, York County, VA	Superfund Mining site	Ground water and surface water were contaminated with heavy metals from the disposal of fly ash. The cleanup plan eliminated contact with the fly ash and contaminated water, restored ground water, and protected nearby wetlands.	The site is being reused as a recreational complex, including ponds and the County Memorial Tree Grove. The site cleanup also protects nearby ponds, a creek, and an estuary, and it is part of a large water quality improvement that has led to the reopening of the Chisman Creek estuary for private and commercial fishing.	Not specified	Not specified	Andrew C. Palestini EPA Region 3 1650 Arch Street Mail Code: 3HS23 Philadelphia, PA 19103-2029 215-814-3233 palestini.andrew@epa.gov	http://www.epa.gov/superfund/programs/recycle/live/casestudy chisman.html
College Park Landfill, Beltsville, MD	Superfund Landfill	Remediation included installing a cap over a landfill that accepted household trash,	The vegetative cover will include diverse native plantings.	The stakeholders were concerned about whether the vegetation would be killed by methane from the	A pilot study is being conducted to ensure these concerns are addressed.	Karen Zhang, PhD, PE, RPM USDA 10300 Baltimore Avenue Bldg. 003, Rm. 117 Beltsville, MD 20705 301-504-5557 zhangk@ba.ars.usda.gov	http://www.wildlifehc.org/ceweb editpro/items/O57F3070.pdf

(Continued)

Property Name and Location	Property Type	Cleanup Type	Revitalization/Reuse Component	Problems/Issues	Solutions	Point of Contact	Notes/Links*
		as well as commercial, industrial and some agricultural and research waste.		landfill, and if the vegetation would be able to adequately prevent leachate generation.			
Craig Farm Drum, Parker, PA	Superfund Landfill	Ground water and soil were contaminated with resorcinol and VOCs, such as benzene and toluene. Site remediation consisted of excavating and stabilizing contaminated soils onsite from two former waste disposal pits.	Wetlands were built on site to replace a smaller area of wetlands lost during construction of the on-site landfill.	Not specified	Not specified	John Epps EPA Region 3 1650 Arch Street Mail Code: 3HS33 Philadelphia, PA 19103-2029 215-814-3144 epps.john@epa.gov	*http://www.epa. gov /reg3hscd/s* uper/sites/PAD 980508527/
DeSale Restoration, Butler County, PA	Pennsylvania Department of Environmental Protection Mining Site	A passive treatment system was used to capture and treat acid mine drainage	In addition to creating a treatment wetland complex, 11 miles of streams that were once devoid of	Not specified	Not specified	Scott Roberts Pennsylvania Deparament of Environmental Protection Office of Mineral Resources	http://www.srwc. org/projects/d esale.php

(Continued)

Property Name and Location	Property Type	Cleanup Type	Revitalization/Reuse Component	Problems/Issues	Solutions	Point of Contact	Notes/Links*
		and included an anoxic collection system, vertical flow ponds, a settling pond and wetland complex, and horizontal flow limestone bed.	life because of acid mine drainage are now teeming with fish.		-	P.O. Box 2063 Harrisburg, PA 171 05-2063 717-783-5338 jayroberts@state.pa.us	
E.I. DuPont Nemours & Co., Inc. (Newport Pigment Plant Landfill), Newport, DE	Superfund Landfill	Soils, sediments, ground water, and surface water were contaminated with various metals. Contaminated sediments were excavated, the two landfills were capped, and soil at the ballpark was removed	The cleanup is protecting Delaware's natural resources and wildlife habitat. Over 35 acres of wetlands and wildlife habitat have been restored as part of the site's overall cleanup.	Ground water appeared to be seeping over the sheet pile wall in several areas of the north landfill. This created a concern regarding possible vapor intrusion into structures above the contaminated ground water plume.	Evaluation of vapor intrusion potential and appropriate mitigation steps was conducted. Ground water table elevation at the north landfill was continuously monitored; water, soil and/or sediment sampling was conducted; and the need for more	Randy Sturgeon EPA Region 3 1650 Arch Street Mail Code: 3HS23 Philadelphia, PA 19103-2029 215-814-3227 sturgeon.randy@epa.gov	http://www.epa.gov/superfund/sites/fiveyear/f0503006.pdf

(Continued)

Property Name and Location	Property Type	Cleanup Type	Revitalization/Reuse Component	Problems/Issues	Solutions	Point of Contact	Notes/Links*
					recovery wells was evaluated		
Former Elf Atochem North America (Bensalem Redevelopment), Cornwell Heights, PA	RCRA Corrective Action Manufacturing Facility Refinery	Site soils and ground water are contaminated with chlorinated organics, PAHs, PCBs, pesticides, and arsenic. Remediation included removing contaminated soil and reusing concrete from demolished buildings as fill for basement areas in buildings that had been razed.	The site is planned to be redeveloped as a mixed-use area with greenspace for passive and active recreation along the Delaware River waterfront.	The property is in an area where many industries have downsized or discontinued operations over the last 20 years. Unemployment rates in the area are among the highest in Bucks County.	The redevelopment authority received a grant and loan from the Brownfields Program to help with the cost of the cleanup. A mixed-use area is planned for the site.	Andrew Clibanoff EPA Region 3 1650 Arch Street Mail Code: 3WC22 Philadelphia, PA 19103-2029 215-814-3391 clibanoff.andrew@epa.gov	http://www.epa.gov/reg3wcmd/ca/pdf/elfato-chem.pdf
Grace Lease Property, Lancaster	Brownfields	A Phase II Environmental Site Assessment found that no	The area, previously abandoned and unused, now ecreational greenspace with	Site remediation was not necessary.	Not applicable	Andrew Kreider EPA Region 3 1650 Arch Street Mail Code: 3HS51 Philadelphia, P	http://www.epa.gov/region03/revitalization/newsletter/spring07/Lorax.html

(Continued)

Property Name and Location	Property Type	Cleanup Type	Revitalization/Reuse Component	Problems/Issues	Solutions	Point of Contact	Notes/Links*
		contaminants were present at levels above state standards, so cleanup was not necessary.	hiking trails, picnic grounds, and a scenic overlook of the Susquehanna River. In addition, Bald Eagle nesting sites have reemerged on the land.			A 19103-2029 215-814-3301 kreider.andrew@epa.gov	
GSA Southeast Federal Center, Washington D.C.	RCRA Corrective Action Manufacturing Facility	Contamination resulted from shipbuilding and ordnance production activities. Eleven of the 14 buildings were deconta-minated and demolished; the remaining buildings will be renovated and reused. Contaminated soil was removed, and ground water is being treated to break down gasoline constituents.	Revitalization includes developing a waterfront park that includes wildlife habitat.	Not specified	Not specified	Barbara Smith EPA Region 3 1650 Arch Street Mail Code: 3LC20 Philadelphia, PA 19103-2029 215-814-5786 smith.barbara@epa.gov	http://www.epa.gov/reg3wcmd/ca/dc/pdf/dc8470090004.pdf

(Continued)

Property Name and Location	Property Type	Cleanup Type	Revitalization/Reuse Component	Problems/Issues	Solutions	Point of Contact	Notes/Links*
Honeywell (Formerly Allied Signal) Baltimore Works Facility, Baltimore, MD	RCRA Corrective Action Industrial Facility	Manufacturing buildings and associated hazardous waste were removed. The containment area was surrounded by a slurry wall and capped, and ground water is being pumped and treated off site. Chromium and PAH-contaminated soil was removed.	A waterfront park will be constructed and is planned to include wildlife habitat.	Not specified	Not specified	Russell Fish EPA Region 3 1650 Arch Street Mail Code: 3LC20 Philadelphia, PA 19103-2029 215-814-3226 fish.	http://www.epa.gov/reg3wcmd/ca/md/pdf/mdd06939671 1 .pdf
Jacks Creek/ Sitkin Smelting & Refining, Inc, Maitland, PA	Superfund Metals Reclam-ation Facility	The former smelting and precious metals reclamation facility contained.	The floodplain remediation required removing vegetation in a segment of the riparian corridor of the creek. Because soil	Not specified	Not specified	Rashmi Mathur, RPM EPA Region 3 1650 Arch Street Mail Code: 3HS22 Philadelphia, PA 19103-2029 215-814-5234 mathur.rashmi@epa.gov	http://www.epa.gov/reg3hwmd/risk/eco/restoration/cs/JacksCreek.htm

(Continued)

Property Name and Location	Property Type	Cleanup Type	Revitalization/Reuse Component	Problems/Issues	Solutions	Point of Contact	Notes/Links*
		several buildings, waste piles, and large areas of soil contaminated with lead, copper, zinc, cadmium, and PCBs. Floodplain wetlands on site and Jacks Creek sediment near the site were contaminated with runoff from the waste piles and soil. The cleanup involved dredging contaminated sediment from the adjacent Jacks Creek,	excavation affected existing wetlands on site, wetlands were recreated in the riparian corridor along Jacks Creek. RPMs created vernal pools, placed woody debris in the wetland as invertebrate habitat, and used a wet meadow seed mix. A monitoring plan will help document the effectiveness of the created wetland.				

(Continued)

Property Name and Location	Property Type	Cleanup Type	Revitalization/Reuse Component	Problems/Issues	Solutions	Point of Contact	Notes/Links*
		excavating contaminated soil, and removing USTss and drums. Contaminated soil, sediment, and waste piles were consolidated and capped. Drums and waste were removed from the site.					
Hopewell Plant (Honey-well), Hopewell, VA	RCRA Corrective Action Manufacturing Facility	This industrial chemical and fertilizer manufacturing facility is being cleaned up to control ground water releases and current human and ecological exposure to contaminated media.	A portion of the facility has been converted to a wildlife habitat area and has been certified as such by the Wildlife Habitat Council.	Not specified	Not specified	Russell Fish EPA Region 3 1650 Arch Street Mail Code: 3LC20 Philadelphia, PA 19103-2029 215-814-3226 fish.	http://www.wildlifehc.org/RegistryCertifiedSites/certsitesdetail2.cfm?LinkAdvID=95327

(Continued)

Property Name and Location	Property Type	Cleanup Type	Revitalization/Reuse Component	Problems/Issues	Solutions	Point of Contact	Notes/Links*
Mill Creek Dump, Erie, PA	Superfund Landfill	A former freshwater wetland that was used as a landfill for foundry sands, solvents, waste oils, and other industrial and municipal waste was capped and flatter slopes were created.	The former landfill is now a golf course. Eight acres of wetlands were constructed adjacent to the course.	Not specified	Not specified	Romuald A. Roman, RPM EPA Region 3 1650 Arch Street Mail Code: 3HS22 Philadelphia, PA 19103-2029 215-814-3212 roman.romuald@epa.gov	http://www.epa.gov/reg3hscd/npl/PAD980231690.htm
Morgantown Ordnance Works Disposal Area - OU1, Monongalia County, WV	Superfund Chemical Production Facility Landfill	Remediation activities included constructing a cap, removing soil and sediment contaminated with heavy metals and PAHs, and constructing three wetlands.	Wetlands were constructed and provided leachate treatment.	Contaminated sediment and soil were intended to be cleaned through bioremediation. However, bioremediation did not meet the clean up standards within a reasonable time frame and was not cost effective.	Three consecutive treatment wetlands were constructed to treat landfill leachate. Monitoring was implemented to ensure the effectiveness of wetlands.	Mr. Hilary Thornton, RPM EPA Region 3 1650 Arch Street Mail Code: 3HS23 Philadelphia, PA 19103-2029 215-814-3323 thornton.hilary@epa.gov	http://epa.gov/reg3hwmd/npl/WVD000850404.htm

(Continued)

Property Name and Location	Property Type	Cleanup Type	Revitalization/Reuse Component	Problems/Issues	Solutions	Point of Contact	Notes/Links*
Naval Amphibious Base Little Creek, Virginia Beach, VA	Superfund Landfill	Approxi-mately 29,000 tons of non-hazardous soil and debris were removed from the landfill and 6,300 cubic yards of clean fill were imported.	The landfill was converted to a tidal wetland. Two connecting channels were constructed to allow tidal inundation into the site from Little Creek Cove. Plants were placed along designated elevations to establish tidal wetland vegetation, using the neighboring marsh as a reference.	Not specified	Not specified	Bruce Pluta EPA Region 3 1650 Arch Street Mail Code: 3HS41 Philadelphia, PA 19103-2029 215-814-2380 pluta.bruce@epa.gov	http://public.lantops-ir.org/sites/public/nable/Site%20Files/IRhistory.aspx#Site%208
Ohio River Park, Neville Island, PA	Superfund Landfill	A previous municipal landfill operating from the 1930s until the 1 950s was capped with a protective cover.	The site will be transformed into a sports complex, with areas of habitat for wildlife; visitors will also be able to enjoy numerous walking, hiking, and biking trails.	Not specified	Not specified	Romuald A. Roman, RPM EPA Region 3 1650 Arch Street Mail Code: 3HS22 Philadelphia, PA 19103-2029 215-814-3212 roman.romuald@epa.gov	http://www.epa.gov/superfund/programs/recycle/live/casestudyohioriver.html
Palmerton Zinc Pile Superfund Site,Palmer-ton, PA	Superfund Mining Site	Former smelting operations resulted in soil and shallow	For the Blue Mountain revegetation, site managers constructed a self-sustaining	Attempting to establish forestland at the site was extremely challenging	Forestland was ultimately abandoned in favor of meadowland.	Charlie Root, RPM EPA Region 3 1650 Arch Street Mail Code: 3HS21	http://costperformance.org/pdf/20070522396.pdf

(Continued)

Property Name and Location	Property Type	Cleanup Type	Revitalization/Reuse Component	Problems/Issues	Solutions	Point of Contact	Notes/Links*
		ground water contamination by heavy metals, such as lead, cadmium, and zinc, and created a defoliated area on the adjacent Blue Mountain, a cinder bank, and additional defoliation along Stoney Ridge. Heavy metals were being transported to nearby stream segments through erosion. Biosolids were applied to accelerate revegetation of the defoliated areas, to stabilize the area, reduce soil erosion caused by wind and surface water, and	meadowland because of minimum metal uptake from the plants. Also, ree species with high metal uptake were removed. For the cinder bank revegetation, the team used a grass seed mixture that included a nitrogen-fixing legume to maintain nitrogen fertility without the need for fertilizer.	because of competition from grasses, animal grazing, and insects. Some grass species were not desirable because of metals uptake. Use of sludge as a soil amendment caused a negative public perception.	The types of grass seeds were replaced with those having minimal metals uptake. Sludge application was replaced with mushroom compost.	Philadelphia, PA 19103-2029 215-814-3193 root.charlie@epa.gov	

(Continued)

Property Name and Location	Property Type	Cleanup Type	Revitalization/Reuse Component	Problems/Issues	Solutions	Point of Contact	Notes/Links*
		increase evapotranspiration to prevent percolation of water and contaminants to the ground water. In addition, a system was installed to divert surface water around the cinder bank and treat leachate before discharge to the creek.					
Resin Disposal, Jefferson Borough, PA	Superfund Landfill	The landfill, which accepted industrial waste including benzene and toluene, was covered with multi-layer cap. Leachate was collected and separated, and oil	The site now contains native wild flowers and is habitat to migratory birds.	Not specified	Not specified	Rashmi Mathur, RPM EPA Region 3 1650 Arch Street Mail Code: 3HS22 Philadelphia, PA 19103-2029 215-814-5234 mathur.rashmi@epa.gov	http://cfpub.epa.gov/supercpad/cursites/csitinfo.cfm?id=0301042

(Continued)

Property Name and Location	Property Type	Cleanup Type	Revitalization/Reuse Component	Problems/Issues	Solutions	Point of Contact	Notes/Links*
		was recycled as fuel for a nearby plant.					
Revere Chemical, Nockamixon Township, PA	Superfund Waste Processing Facility	The site was contaminated with benzoic acid, VOCs, solvents, and PAHs. Remediation included disposing of debris and solid wastes off-site, cleaning VOC-contaminated soil by vacuum extraction, and installing a slurry wall and cap over an area contaminated with hazardous waste associated with an acid and metal-plating waste processing facility.	Revitalization activities included planting wildflowers and other foliage to attract migratory birds and other wildlife.	Treatment of VOC-contaminated soil by in situ vacuum extraction did not meet requirements of the Pennsylvania Land Recycling and Remediation Standards Act.	Protective levels of contaminant concentrations in ground water were established usingthe Synthetic Precipitation Leaching Procedure to determine the extent of capping. Soil contaminated with VOCs was treated by ex situ vacuum extraction.	Melissa Friedland EPA HQ Ariel Rios Building 1200 Pennsylvania Avenue Mail Code: 5204P Washington, DC 20460 703-603-8864 friedland.melissa@epa.gov	http://cfpub.epa.gov/supercpad/cursites/csitinfo.cfm?id=0300982

(Continued)

Property Name and Location	Property Type	Cleanup Type	Revitalization/Reuse Component	Problems/Issues	Solutions	Point of Contact	Notes/Links*
Saltville Waste Disposal Ponds, Saltville, VA	Superfund Manufacturing Facility	Elevated mercury levels were present in soil and ground water in the area beneath the former chlorine plant. Remediation activities included constructing a water treatment plant and capping the ponds.	A wildlife habitat area was created on the former disposal ponds.	Not specified	Not specified	Eric Newman 1650 Arch Street Mail Code: 3HS23 Philadelphia, PA 19103-2029 215-814-3237 newman.eric@epa.gov	http://www.epa.gov/reg3hscd/super/sites/VAD003127578/
Seaford Nylon Plant, Seaford, DE	RCRA Corrective Action Site Manufacturing Facility	Wastes include fly ash, corrosives, ignitables, spent halogenated solvents, and discarded commercial chemical products. Ground water contains low levels of metals and VOCs	Reuse includes expansion of the neighboring golf course.	There was concern that the fly ash placed at the golf course may cause a ground water problem.	Evaluations of the ground water at the golf course indicated that the fly ash did not impact the ground water.	Douglas Zeiters Delaware Department of Natural Resources and Environmental Control 89 Kings Highway Dover, DE 19901 302-739-9403 douglas.zeiters@state.de.us	http://www.epa.gov/reg3wcmd/ca/de/pdf/ded002348845.pdf

(Continued)

Property Name and Location	Property Type	Cleanup Type	Revitalization/Reuse Component	Problems/Issues	Solutions	Point of Contact	Notes/Links*
		and low pH. Remediation included MNA of ground water with ICs as well as installing a protective cover over solid waste. Fly ash from the site was used as fill at an adjacent golf course.					
Site 46 Landfill A, Stump Dump Road, Dahlgren, VA	Superfund Landfill	Ground water and surface water contained contaminants such as cadmium, lead, mercury, and PCBs from municipal waste at the site. Contaminated waste from the site was removed to an appropriate off-site landfill.	The remedial design includes the integration and establishment of tidal wetlands in the low areas of the site.	Uncovering UXO caused a safety issue at the site.	EOD support and screening at all times was required.	Neal Parker 1314 Harwood St., SE Washington Navy Yard Washington, D.C. 20374 202-685-3281 parkernm@efaches .navfac.navy.mil	http://www. wildlifehc. org/eweb editpro/items/ O57F3079.pdf

(Continued)

Property Name and Location	Property Type	Cleanup Type	Revitalization/Reuse Component	Problems/Issues	Solutions	Point of Contact	Notes/Links*
Tybouts Corner Landfill, New Castle, DE	Superfund Landfill	Remediation activities included installing water lines for residents in the area and installing a protective cap over the landfill, which accepted municipal and household waste.	Revitalization included planting wildflowers and other vegetation on the cap to stabilize the ground and prevent erosion.	Not specified	Not specified	Katherine Lose, RPM EPA Region 3 1650 Arch Street Mail Code: 3HS23 Philadelphia, PA 19103-2029 215-814-3240 lose kate@epa.gov	http://cfpub.epa.gov/supercpad/cursites/csitinfo. cfm?id=0300035
Walsh Landfill, PA	Superfund Landfill	Residential well water off-site was contaminated with chloromethane, chloroform, xylenes, and other VOCs, as well as lead, mercury, and zinc. Remediation included removing waste and installing an evapotrans-piration cover	Revitalization included replanting a vegetative layer of a variety of native hardwood and coniferous trees.	The site was planned for reuse originally. However, because both the site owner and community were unresponsive, the team installed an evapotranspiration cover with trees as an integral part of the remedy. Therefore, reuse options are minimal.	Trees planted as the vegetative layer of the evapotranspiration cover have provided excellent habitat for birds and small mammals. Current plans are for the site to remain as is.	Frank Klanchar, RPM EPA Region 3 1650 Arch Street Mail Code: 3HS22 Philadelphia, PA 19103-2029 215-814-3218 klanchar.frank@epa.gov	http://www.epa.gov/reg3hwmd/super/sites/PAD980829527/index.htm

(Continued)

Property Name and Location	Property Type	Cleanup Type	Revitalization/Reuse Component	Problems/Issues	Solutions	Point of Contact	Notes/Links*
		system to protect against migration of on site ground water contaminated with mercury, toluene, and other VOCs from former disposal practices.					
Wildcat Landfill, Dover, DE	Superfund Landfill	Contaminated soil and ground water from the previous landfill were capped with a protective cover.	A mixture of native plants and wildflowers were planted on the cap, and Kent County is evaluating plans to allocate a part of the site as a greenway, which is an open space for recreational purposes.	Not specified	Not specified	Hilary Thornton EPA Region 3 1650 Arch Street Mail Code: 3HS23 Philadelphia, PA 19103-2029 215-814-3323 thornton.hilary@epa.gov	http://cfpub.epa.gov/supercpad/cursites/csitinfo.cfm?id=0300101
Wood-lawn County Landfill, MD	Superfund Landfill	The ground water is contaminated with VOCs, primarily vinyl chloride and 1,2-dichloroethane, and with PAHs,	The closed landfill was used to create wildlife habitat called "New Beginnings, the Woodlawn Wildlife Habitat Area." It is currently used as a	Analyses showed contamination of on-site and off-site ground water, soil, and sediment and surface water of a	The original remedy included extraction and treatment of contaminated	James J. Feeney, RPM EPA Region 3 1650 Arch Street Mail Code: 3HS22 Philadelphia, PA 19103-2029 215-814-3190 feeney.jim@epa.gov	http://www.wildlifehc.org/brownfields/woodlawn.cfm

(Continued)

Property Name and Location	Property Type	Cleanup Type	Revitalization/Reuse Component	Problems/Issues	Solutions	Point of Contact	Notes/Links*
		pesticides, and metals, primarily manganese. Initially RPMs installed an impermeable cap and ground water P&T system. Later they replaced the cap with a vegetative soil cap to help sustain naturally occurring bacteria in the soil that degrade the contaminants. In addition to P&T, the remedy included MNA with monitoring of the ground water and the vegetative soil cover. The team planted wildlife enhancements such as trees and native	nature and science study area by local schools and as an area for projects by the Boy Scouts and Girls Scouts of America.	stream that crosses the site. MNA posed a difficulty due the scarcity of its use at the time.	ground water. However, continued monitoring showed that MNA effectively removed or immobilized contaminants from ground water. Two remedial designs were completed in parallel in case the MNA process failed to perform as expected.		

(Continued)

Property Name and Location	Property Type	Cleanup Type	Revitalization/Reuse Component	Problems/Issues	Solutions	Point of Contact	Notes/Links*
		wildflowers after installing the vegetative cap.					
REGION 4							
Black Warrior-Cahaba Rivers Land Trust, AL	Brownfields Mining Site	Soils contaminated with lead and heavy metals. Remediation included a recreational park and community stream cleanup events.	Transformed a former industrial region into a 27-mile greenway with parks and paths along the Five-Mile Creek.	It could take 20 years to complete the entire greenway project.	Many of the targeted former industrial areas have been cleaned up and made available to communities as natural and recreational land.	EPA Region 4 Brownfields Team 61 Forsyth Street, S.W. Atlanta, GA 30303-8960 404-562-8493 www.epa.gov/region4/waste htm	http://www.epa.gov/brownfields/success/fultondale al BRA G.pdf
Milan Army Ammunition Plant, Milan, TN	Superfund Ammunitions Plant	Two wetland systems were created, a subsurface flow ground-bed wetland and a surface flow lagoon wetland, to degrade explosives and their byproducts. Specifically,	Revitalization included creation of wetlands and use of phytoremediation as a remedial technology.	Weather was an obstacle because it affects the efficiency of phytoremediation.	Not specified	Laurie Haines U.S. Army Environmental Center 2511 Jefferson Davis Highway Taylor Building NC3- Arlington, VA 22202-3926 703-601-1590 laurie.haines@us.army.mil	http://www.wildlifehc.org/cweb editpro/items/O57F3081 .pdf

(Continued)

Property Name and Location	Property Type	Cleanup Type	Revitalization/Reuse Component	Problems/Issues	Solutions	Point of Contact	Notes/Links*
		ground water was contaminated with explosives constituents including TNT, RDX, HMX, 2,4-DNT and 2,6-DNT.					
Northwest 58th Street Landfill, Miami, FL	Superfund Landfill	Ground water contaminated with heavy metals and toxic chemicals from previous landfill activities was cleaned up through remediation and closure of the landfill.	Through careful design, a lake was constructed at the site for wading birds; trails were created with lookout centers.	Not specified	Not specified	Bill Denman EPA Region 4 61 Forsyth Street, SW Atlanta, GA 30303 404-562-8939 denman.bill@epa.gov	http://www.epa.gov/region4/waste/reuse/fl/nw58reuse.pdf
Solitron Microwave, Port Salerno, FL	Superfund Manufacturing Facility	Ground water contaminants consist of PCE and its breakdown products. Remediation activities include water line extensions, soil removal, *in situ* chemical oxidation, and natural attenuation.	Six acres at the site have been reserved for wetland areas, an upland preserve for native plant habitat, and a 50-foot natural buffer between the site and surrounding residential areas.	Not specified	Not specified	Bill Denman EPA Region 4 61 Forsyth Street, SW Atlanta, GA 30303 404-562-8939 denman .bill@epa.gov	http://www.epa.gov/Region4/waste/npl/nplfls/solmicfl.htm

(Continued)

Property Name and Location	Property Type	Cleanup Type	Revitalization/Reuse Component	Problems/Issues	Solutions	Point of Contact	Notes/Links*
REGION 5							
Allied Chemical & Ironton Coke, Ironton, OH	Superfund Chemical and Tar Manufacturing Facility	Solid wastes and wastewater including crude tar and ammonia contaminated the ground water at this site. Remediation activities included excavating and disposing of contaminated soil, installing containment systems, and constructing a water treatment plant.	This area is being converted into a wetlands system, taking advantage of its natural flooding conditions and predisposition to wetlands- type vegetation.	Not specified	Not specified	Syed Quadri EPA Region 5 77 West Jackson Boulevard Mail Code: SR-6J Chicago, IL 60604-3507 312-886-5736 quadri.syed@epa.gov	http://www.epa.gov/region5/sit es/alliedchemical/pdfs/allied-chemical-5yr-review-200409-report.pdf
Bowers Landfill, Circleville, OH	Superfund Landfill	Soil, ground water, and surface water contaminated with VOCs and PCBs. Remediation included removing debris and installing a clay cap.	Wetlands were created around the site to protect the cap from flooding.	The nearby Scioto River was prone to flooding, which could affect the landfill cap.	Wetlands were created in the area between the landfill and river, where clay was taken to create the cap, to control flooding.	Sirtaj Ahmed, RPM EPA Region 5 77 West Jackson Boulevard Chicago, IL 60604-3507 312-886-4445 ahmed.sirtaj@epa.gov	http://www.epa.gov/superfund/programs/recycle/live/casestudybowers.html

(Continued)

Property Name and Location	Property Type	Cleanup Type	Revitalization/Reuse Component	Problems/Issues	Solutions	Point of Contact	Notes/Links*
Calumet Container Site, Hammond, IN	Superfund Industrial Facility	Remediation consisted of cleaning up soil contamination caused by previous drum and pail reconditioning operations at the site.	The area will be restored as a native habitat area with opportunities for passive recreation, including walking trails, and increasing biological diversity of native plants for prairie and wetland habitats.	Not specified	Not specified	Thomas Bloom EPA Region 5 77 West Jackson Boulevard Mail Code: SE-4J Chicago, IL 60604-3507312-886-1967 bloom.thomas@epa.gov	http://www.epa.gov/region5superfund/redevelop/pdf/Calumet.pdf
Broverman Landfill, Christian County, IL	Illinois EPA Corrective Action Landfill	Cleanup included repair of the protective cap placed over an abandoned municipal landfill.	Prairie plants were seeded to stabilize the soil cover and reduce maintenance requirements.	Deep gullies were eroding down the landfill's sparsely vegetated sides and low areas were holding pools of stagnant water.	The cleanup team filled in large surface irregularities, added rip-rap in drainage ways to deter future erosion, installed vegetation mats, and seeded the area with native grasses and wildflowers	Jody Kershaw Illinois EPA 1021 North Grand Avenue East P.O. Box 19276 Springfield, Illinois 62794-9276 217-524-3285 jody.kershaw@epa.state.il.us	http://www.epa.state.il.us/environmental-progress/v25/n1/abandoned-landfill.html

(Continued)

Property Name and Location	Property Type	Cleanup Type	Revitalization/Reuse Component	Problems/Issues	Solutions	Point of Contact	Notes/Links*
					The remedy was cost-effective because nitrogen and phosphorous did not have to be added to the soil, additional topsoil and tilling was not required, and maintenance only included occasional prescribed burns.		
Dupage County Landfill, IL	Superfund Landfill	Ground water contamination associated with the landfill was cleaned up.	The site is now being used as a recreational area with picnic and camping areas, trails, and a lake. The previous landfill is used for sledding during the winter months.	Not specified	Not specified	Thomas Williams, RPM EPA Region 5 77 West Jackson Boulevard Mail Code: SR-6J Chicago, IL 60604-3507 312-886-6157 williams.thomas@epa.gov	http://cfpub.epa.gov/supercpad/cursites/csitinfo.cfm?id=0500606

(Continued)

Property Name and Location	Property Type	Cleanup Type	Revitalization/Reuse Component	Problems/Issues	Solutions	Point of Contact	Notes/Links*
E-Pond Solid Waste Management Unit, Lima, OH	RCRA Corrective Action Refinery Landfill	Synthetic root barrier and soil cover will be placed over the site, which is contaminated with chromium, antimony, thallium, PCB-1248, benzo(a)pyrene, and dibenz(a,h)anthracene.	Prairie habitat constructed with native plants.Interpretive areas and educational opportunities will be created.	Not specified	Not specified	Thomas Matheson, RPM EPA Region 5 77 West Jackson Boulevard Mail Code: DM-7J Chicago, IL 60604-3507 312-886-7569 matheson.thomas@epa.gov	http://www.epa.gov/epaoswer/hazwaste/ca/curriculum/download/eco-rec.pdf
Fernald, Southwest OH	Superfund Uranium Metal Production	Remediation and closure project addressing uranium contamination in soil and ground water. Remediation included treatment and disposal through an on-site disposal facility and off-site disposal. The treated silos and waste pit materials were all disposed of off-site. The on-site disposal facility contains primarily contaminated soil and building debris.	End use of the entire 1,000- acre site is an educational park focusing on site history and ecology. Deep excavations are being converted to wetland and open water habitat. Excavations into the subsoil are being converted to native grasslands.	The primary problems have been invasive species control, geese and deer browsing, and germination success.	Invasive control was initially implemented through mechanical removal. Selective use of herbicides provides on-going control. Deer exclosures have been installed to fence the deer out of new restoration areas where		

(Continued)

Property Name and Location	Property Type	Cleanup Type	Revitalization/Reuse Component	Problems/Issues	Solutions	Point of Contact	Notes/Links*
					woody plants were installed. Goose fencing, flagged twine, and coyote decoys have been used to discourage geese. Germination success is being evaluated and in some cases has required reseeding.	Thomas A. Schneider Ohio EPA, Office of Federal Facility 401 East Fifth Street Dayton, OH 45402-2911 937-285-6466 tom.schneider@epa.state.oh.us	http://www.wildlifehc.org/cweb editpro/items/O57F3069.pdf
Ford Rouge Center, Dearborn, MI	MDEQ/ RCRA Corrective Action Automobile Manufacturing Complex	Remediation included removal of soils contaminated with SVOCs, PCBs, metals, and organics as well as containment strategies.	Ecological enhancements include a vegetated roof, pervious pavement, vegetated drainage swales, hedgerow wildlife corridors, wetland restoration, sunflower	Issues encountered included coordinating remediation with ongoing plant expansion activities.	Early negotiations with MDEQ helped the process go smoothly.	Dan Ballnik Ford Motor Company One American Road Dearborn, MI 48126 313-248-8606 dballni1@ford.com	http://www.wildlifehc.org/cweb editpro/items/O57F3071 .pdf

(Continued)

Property Name and Location	Property Type	Cleanup Type	Revitalization/Reuse Component	Problems/Issues	Solutions	Point of Contact	Notes/Links*
			plantings, and grassland restoration. When it was built, this was the world's largest green roof at 10 acres in size. Honey bee hives have been added to enhance pollination for new plantings.				
Former Brass Foundry and Eljer Park, Marysville, OH	RCRA Corrective Action Foundry	Remediation included removing soil and stream sediments contaminated with VOCs and metals, demolishing buildings, capping residual areas, and improving site drainage to prevent erosion.	Revitalization included creating a park with athletic fields, playground equipment, a walking trail, and a wetlands area.	Not specified	Not specified	Jan J. Chizzonite, Managing Executive Partner Environmental Strategies Consulting LLC 11911 Freedom Drive Reston, VA 20190 703-709-6500 jan.chizzonite@wspgroup.com	http://www.cpa.gov/nc/national caconf/docs/ Chizzonite.pdf
Former Ford Michigan Casting Center Landfill, Flat Rock, MI	Brownfields Landfill	A wooded leachate collection/management system was used to treat contaminated soil and ground water.	Wooded phytoremediation area providing increased biodiversity via creation of wildlife habitat for various birds and small mammals.	Not specified	Not specified	Jeff Hartlund Ford Motor Company One American Road Dearborn, MI 48126 313-322-0700 jhartlun@ford.com	http://www.wildlifehc.org/cweb editpro/items/ O57F3059.pdf

(Continued)

Property Name and Location	Property Type	Cleanup Type	Revitalization/Reuse Component	Problems/Issues	Solutions	Point of Contact	Notes/Links*
Former Gulf Refinery Site, Hooven, OH	RCRA Corrective Action Refinery	Phytoremediation consisting of a vegetative cap was used to treat soil contaminated with a mixture of petroleum hydrocarbons, including PAHs.	Activities at the site include constructing a wetland habitat for wildlife and extending the park planned for the adjacent area by providing community access.	Not specified	Not specified	Lucinda Jackson ChevronTexaco Corporation 100 Chevron Way P.O. Box 1627 Richmond, CA 94802-0627 510-242-1047 luaj@chevron.com	http://www.wildlifehc.org/eweb editpro/items/O57F3061 .pdf
Ilada Energy Company, East Cape Girardeau, IL	Superfund Waste Oil Reclamation Facility	Water and soil were contaminated with VOCs, PCBs, and heavy metals. Remediation activities included the removal of 1,742 cubic yards of soil and 865,700 gallons of water. Oil and sludge were incinerated.	The site is part of an ecological preservation area. The Land Conservancy bought land around the site and planted bottomwood trees adjacent to the site.	Not specified	Not specified	Sam Chummar EPA Region 5 77 West Jackson Boulevard Mail Code: SR-6J Chicago, IL 60604-3507 312-886-1434 chummar.sam@epa.gov	http://www.epa.gov/region5superfund/npl/illinois/ILD980996789.htm
Industrial Excess Landfill (IEL), Uniontown, OH	Superfund Landfill	Remediation activities such as extraction and treatment, capping the landfill, and installing a landfill gas extraction system were used to treat ground water contaminated by VOCs.	The site's remedy involves enhancing wildlife habitat and creating greenspace. Almost 10,000 native trees and shrubs were planted.	Not specified	Not specified	Timothy Fischer, RPM EPA Region 5 77 West Jackson Boulevard Mail Code: SR-6J Chicago, IL 60604-3507 312-886-5787 fischer.timothy@epa.gov	http://www.epa.gov/superfund/sites/fiveyear/f2006050001133 .pdf

(Continued)

Property Name and Location	Property Type	Cleanup Type	Revitalization/Reuse Component	Problems/Issues	Solutions	Point of Contact	Notes/Links*
Joliet Army Ammunition Plant, Joliet, IL	Superfund Ammunitions Plant	Remediation included excavation and off-site disposal of soils contaminated with metals and on-site bioremediation of explosives-contaminated soils.	Midewin National Tall Grass Prairie was created for recreational, educational, and agricultural benefits to the public. Also, revitalization activities included restoring native wildlife populations and habitat.	Remediation goals were questioned as possibly not protecting ecological resources of the Midewin National Tall Grass Prairie due to the uncertainty of the risk posed by chemical constituents.	Site representatives are still working to establish proper remediation goals and costs.	Laurie Haines U.S. Army Environmental Center 2511 Jefferson Davis Highway Taylor Building NC3- Arlington, VA 22202-3926 703-601-1590 laurie.haines@hqda.army.mil	http://www.epa.gov/R5Super/npl/illinois/IL0210090049.htm
Petersen Sand and Gravel, Libertyville, IL	Superfund Quarry	The former Petersen quarry was used during the 1950s as a dumping ground for solvents and paints causing extensive contamination. Cleanup activities included removing drums, paint cans, and contaminated soil and surface water.	The cleanup enabled Independence Grove Forest Preserve to create a 115-acre lake and establish an education center at the site.	Not specified	Not specified	David Seeley, RPM EPA Region 5 77 West Jackson Boulevard Mail Code: SR-6J Chicago, IL 60604-3507 312-886-7058 seely.david@epa.gov	http://www.epa.gov/region5superfund/npl/illinois/ILD003817137.htm

(Continued)

Property Name and Location	Property Type	Cleanup Type	Revitalization/Reuse Component	Problems/Issues	Solutions	Point of Contact	Notes/Links*
Pocket Parks at Former Service Stations, Chicago, IL	IEPA Corrective Action Former Service Station	The sites were contaminated with BTEX, and contaminated soil was removed. Each of the sites received "No Further Remediation" letters through IEPA's Voluntary Cleanup Program.	Greenspace was created to reduce paved areas, which decreased the amount of stormwater that reaches the combined storm sewers.	Local politics favored commercial use over recreational use.	Multiple meetings with community groups helped to achieve consensus.	Kelly Kennoy City of Chicago 30 North Lasalle Street, 25th Floor Chicago, IL 60602-2575 312-744-8692 kkennoy@cityofchicago.org	http://www.wildlifehc.org/ eweb editpro/items/ O57F3057.pdf
REGION 6							
AMAX Metals Recovery (Freeport McMoRan), Braithwaite, LA	RCRA Corrective Action Metals Recovery Facility	A UST and waste pile area was cleaned up and designated "ready for reuse."	A water retention pond was dewatered to form a wetland that provided a home to alligators relocated due to Hurricane Katrina in 2005.	Not specified	Not specified	U.S. EPA Region 6 1445 Ross Avenue Suite 1200 Dallas, TX 75202-2733 Louisiana Department of Environmental Quality Galvez Building 602 North Fifth Street Baton Rouge, LA 70802	http://findarticl es. com/p/article s/miqn4200/ is20080604/ai n25483065?tag =artBody;col1
Brooks City-Base, San Antonio, TX	RCRA Corrective Action Former Medical Research and Development Facility	A portion of the base was cleaned up by installing soil vapor extraction and ground water P&T systems, removing and installing a cover over garbage and	The former air force base was issued a "ready for reuse" determination, which was the first of its kind issued in Texas and the first for a federal facility	Not specified	Not specified	Jeanne Schulze EPA Region 6 1445 Ross Avenue, Suite 1200 Mail Code: 6PD-F Dallas, TX 75202-2733 214-665-7254 schulze.jeanne@epa.gov	http://enviro.bl r. com/display.cf m/id/2591 9

(Continued)

Property Name and Location	Property Type	Cleanup Type	Revitalization/Reuse Component	Problems/Issues	Solutions	Point of Contact	Notes/Links*
		construction debris, excavating contaminated soil, and incorporating ICs.	nationwide. The remedial process incorporated ecological revitalization into the cleanup plan.				
DuPont Remington Arms Facility, Lonoke, AK	RCRA Corrective Action Manufacturing Facility	Remediation included excavation and treatment of approximately 6,080 cubic yards of contaminated soils.	Remington Arms continues to manufacture ammunition at the facility. The remaining 731 acres are managed as a wildlife habitat.Ecological revitalization efforts include construction of a 20-acre moist soil impoundment for waterfowl habitat in cooperation with Ducks Unlimited.	Not specified	Not specified	Jeanne Schulze EPA Region 6 1445 Ross Avenue, Suite 1200 Mail Code: 6PD-F Dallas, TX 75202-2733 214-665-7254 schulze.jeanne@epa.gov	http://www.epa.gov/epaoswer/hazwaste/ca/success/rem11-07.pdf
England Air Force Base, LA	RCRA Corrective Action Air Force Base	A portion of the former air force base was cleaned up by removing contaminated soil, incorporating ICs, and	Areas excavated as part of a remedial action became part of the Audubon Trail, providing habitat and a	Not specified	Not specified	Louisiana Department of Environmental Quality Public Records Center Galvez Building, Room 127 602 N. Fifth Street Baton Rouge, LA 70802	http://www.epa.gov/region6/ready4reuse/englandrfr.pdf

(Continued)

Property Name and Location	Property Type	Cleanup Type	Revitalization/Reuse Component	Problems/Issues	Solutions	Point of Contact	Notes/Links*
		instituting MNA of contaminated ground water. The site was designated "ready for reuse."	stopping point for migratory birds, and an expanded 18-hole golf course.				
French, Ltd., Crosby, TX	Superfund Industrial Waste Storage	Remediation included treating soil and ground water contaminated with VOCs and heavy metals and creating 23 acres of new wetlands.	Wetlands and surrounding habitat can be used as recreation for outdoor enthusiasts and as habitat for vegetation and wildlife.	Not specified	Not specified	Ernest Franke, RPM EPA Region 6 1445 Ross Avenue Suite 1200 Mail Code: 6SFRA Dallas, TX 75202-2733 214-665-8521 franke.ernest@epa.gov	http://cfpub.epa.gov/supercpa d/cursites/csitinfo.cfm?id=0602498
Heifer International New World Headquarters, Little Rock, AR	Brownfields Industrial Facility	Petroleum contaminated soil was removed from the site.	Activities at the site included the creation of retention ponds and a wetland habitat.	The primary issue at this site was funding.	Support from federal, state, and local sources, along with existing funds allowed cleanup.	Gerald Cound Director of Facilities Management Heifer International 1 World Avenue Little Rock, AR 72202 501-907-2965 gerald.cound@heifer.org	http://www.wildlifehc.org/ewebeditpro/items/O57F5385.pdf
REGION 7							
3-D Investments, Inc., Alda, NE	RCRA Brownfields and Superfund Former	The 3.65-acre site was investigated under RCRA authority. The facility went bankrupt and cleanup costs	EPA sent a letter stating the facility was cleaned up, and the property was deeded to the Crane	During the cleanup response, EPA discovered areas of contamination	EPA Region 7 RCRA received a RCRA Brownfields	Andrea R. Stone EPA Region 7 901 North Fifth Street Mail Code: ARTDRCAP Kansas City, KS 66101 913-551-7662	http://www.epa.gov/swerosps/rcrabf/html-doc/tsefac03.htm

(Continued)

Property Name and Location	Property Type	Cleanup Type	Revitalization/Reuse Component	Problems/Issues	Solutions	Point of Contact	Notes/Links*
	Gas Station, Battery Cracking and Lead Recovery Facility	exceeded monies in the facility's trust fund, so EPA RCRA referred the facility to Region 7 EPA Superfund. Region 7 Superfund evaluated the site and conducted removal activities of lead-contaminated soils. The site was cleaned up to residential or near residential standards.	Meadows Nature Center, a nonprofit organization dedicated to natural resource education and the preservation of Sandhill cranes.	that were previously unknown. Neighbors and Crane Meadows Nature Center also had a concern regarding excess tree removal.	Prevention Initiative Targeted Site Effort grant to assist with characterization, public involvement and other activities. EPA worked with neighbors and Crane Meadows Nature Center to alleviate their concerns about removing perimeter trees. Crane Meadows Nature Center wanted perimeter	stene.andrear@epa.gov	

(Continued)

Property Name and Location	Property Type	Cleanup Type	Revitalization/Reuse Component	Problems/Issues	Solutions	Point of Contact	Notes/Links*
					trees to remain to serve as a wind-break. EPA obliged this request. Mulch from some of the trees was also left onsite.		
Cherokee County, Galena, KS	Superfund Mining Site	Remediation consisted of burying surface mine wastes contaminated with lead, mercury, and cadmium in abandoned mine pits, subsidence areas, and mine shafts on site; diverting streams away from waste piles; recontouring land surface; and revegetating with native prairie grasses to control runoff and erosion.	Native prairie grassland habitat encouraged the return of wildlife.	Potential for cave-in of filled mine shafts after heavy rain or freezing and thawing cycles.	Avoided development in the areas with potential for cave-in or collapse.	David Drake, RPM EPA Region 7 901 North Fifth Street Mail Code: SUPRFFSE Kansas City, KS 66101 913-551-7626 drake.dave@epa.gov	http://www.epa.gov/superfund/programs/recycle/live/casestudycherokee.html

(Continued)

Property Name and Location	Property Type	Cleanup Type	Revitalization/Reuse Component	Problems/Issues	Solutions	Point of Contact	Notes/Links*
Times Beach, Times Beach, MO	Superfund Contaminated Urban Area	A temporary incinerator was installed to burn soil contaminated with dioxin. The waste ash from the treated soil was buried on site. People were relocated and all homes and businesses were demolished.	A state park now exists on the site and acts as a bird sanctuary.	Numerous problems and issues resulted from this contentious Superfund site. See the Web site provided under "Notes/Links" for more information.	See the Web site provided under "Notes/Links " for more information.	Bob Feild, RPM EPA Region 7 901 North Fifth Street Mail Code: SUPRMOKS Kansas City, KS 66101 913-551-7697 feild.robert@epa.gov	http://cfpub.ep a. gov/supercpa d/cursites/csiti nfo. cfm?id=070 1237
Wheeling Disposal Service Co, Inc. Landfill, Amazonio, MO	Superfund Landfill	Soil contaminated with municipal and industrial wastes was remediated by upgrading the existing landfill cap with a clay and soil cover. Ground and surface water were monitored.	During the cleanup, the owner dug a pond and planted native wild grasses and other foliage that would attract birds and wildlife.	Not specified	Not specified	Amer Safadi, RPM EPA Region 7 901 North Fifth Street Mail Code: SUPRMOKS Kansas City, KS 66101 913-551-7825 safadi.amer@epa.gov	http://cfpub.ep a.gov/ supercpa d/cursites/csiti nfo .cfm?id=070 0780
REGION 8							
BP Former Refinery, Platte River Commons, Casper, WY	RCRA Corrective Action Former Petroleum Refinery	Cleanup included removal of trash and waste from the river to contain the flow of contaminated ground water, excavation of	After the river was cleaned up, a recreational kayak course was created. A portion of the site was used to create an	Not specified	Not specified	Vickie Meredith WDEQ Solid & Hazardous Waste Division, Hazardous Waste Permitting and Corrective Action Program	http://www.ep a. gov/waste ard/correctivea ction/ pdfs/casp er11-07.pdf

(Continued)

Property Name and Location	Property Type	Cleanup Type	Revitalization/Reuse Component	Problems/Issues	Solutions	Point of Contact	Notes/Links*
		contaminated soils, addition of P&T wells and construction of a wetland treatment system. Nearly 2,000 trees were planted to assist with phytoremediation.	18- hole golf course. Wetlands were incorporated into the golf course design to assist in treating contaminated ground water. Trees were planted for phytoremediation.			250 Lincoln Street Lander, WY 82520 vmered@state.wy.us 307-332-6924 Tom Aalto, EPA Region 8 1595 Wynkoop Street Mail Code: 8P-HW Denver, CO 80202-1129 aalto.tom@epa.gov 303-312-6949	
Cache La Poudre River Superfund Site, Fort Collins, CO	Superfund	Soil and sediments in the Poudre River, and ground water were contaminated with gasoline mixed with coal tar. Cleanup activities included sediment excavation and temporary re-routing of the Poudre River, a vertical sheet pile barrier to stop ground water flow, and ground water treatment.	EPA completed an intact but unobtrusive remedy of the Poudre River to preserve the riverine habitat.	Beavers ate about half of the tree plantings.	Site managers used wire on the first 6 to 8 feet of tree plantings, and painted the wire to be easily visible.	Paul Peronard, OSC EPA Region 8 1595 Wynkoop Street Mail Code: 8EPR-SR Denver, CO 80202-1129 303-312-6808 peronard.paul@epa.gov	http://www.clu - in.org/conf/tio /ccocasestudies 080207/

(Continued)

Property Name and Location	Property Type	Cleanup Type	Revitalization/Reuse Component	Problems/Issues	Solutions	Point of Contact	Notes/Links*
California Gulch Superfund Site, Upper Arkansas River Operable Unit, Leadville, CO	Superfund Mining Site	The mining district's soil, surface water, and sediments were heavily contaminated with lead, zinc, and other heavy metals from mine tailings. Biosolids and lime were applied directly to the tailings along Upper Arkansas River.	The area along the river has been restored and supports vegetation and wildlife, and is available for agricultural use and recreational use such as hiking and fishing.	Tailings could not be excavated because of the risk of tailings entering the river and the difficulty of finding a repository for the contaminated soil. Also, replacement of topsoil would be costly. Mobilizing materials to the site was difficult due to the elevation of the site. Water was also scarce due to low rainfall and high elevation.	Biosolids were spread over the tailings, reducing the potential for tailings to migrate to the river.	Rebecca Thomas, RPM EPA Region 8 1595 Wynkoop Street Denver, CO 80202-1129 303-312-6552 thomas.rebecca@epa.gov Mike Holmes, RPM EPA Region 8 1595 Wynkoop Street Denver, CO 80202-1129 303-312-6607 holmes.michael@epa.gov	http://www.ep a. gov/superfund/ programs/recy cle /pdf/calgulc h.pdf
East Helena Site, Helena, MT	Superfund Smelting Site	Ground water, surface water, and soil contamination from decades of lead smelting activities was cleaned up by	In addition to mixed commercial and residential use, portions of the site are being used for a neighborhood park, a	Not specified	Not specified	Scott Brown EPA Region 8 Montana Operations Office Federal Building 10 West 15th Street Suite 3200	http://cfpub.ep a.gov/supercpa d/cursites/csiti nfo.cfm?id=08 0 0377

(Continued)

Property Name and Location	Property Type	Cleanup Type	Revitalization/Reuse Component	Problems/Issues	Solutions	Point of Contact	Notes/Links*
		removing waste, treating soil, and capping the area.	baseball field, and some wetlands redevelopment.			Mail Code: 8MO Helena, MT 59626 406-457-5035 brown.scott@epa.gov	
Kennecott North and South Zone Sites, Salt Lake County, UT	Superfund Mining Site	Soil and ground water were contaminated with mining wastes, including sulfates and heavy metals. Soil was removed, and ground water was pumped and treated in the mine's tailings slurry line.	Open space, wetlands, and wildlife habitat were created. A residential area was also created.	Not specified	Not specified	Rebecca Thomas, RPM EPA Region 8 1595 Wynkoop Street Mail Code: 8EPR-SR Denver, CO 80202-1129 303-312-6552 thomas.rebecca@epa.gov	http://www.epa.gov/superfund/programs/aml/tech/kennecott.pdf
Milltown Reservoir Sediments, Milltown, MT	Superfund Mining Site	Six million cubic yards of mining waste that had piled up at the base of the Milltown Dam was poisoning the reservoir and affecting drinking water. A new drinking water system was installed at the site.	In addition to adding a new drinking water system, 2.5 miles was added to existing hiking trails in Missoula to complete a loop around the University of Montana and Missoula's waterfront.	Not specified	Not specified	Scott Brown EPA Region 8 Montana Operations Office Federal Building 10 West 15th Street Suite 3200 Mail Code: 8MO Helena, MT 59626 406-457-5035 brown.scott@epa.gov	http://cfpub.epa.gov/supercpad/cursites/csitinfo.cfm?id=080 0445

(Continued)

Property Name and Location	Property Type	Cleanup Type	Revitalization/Reuse Component	Problems/Issues	Solutions	Point of Contact	Notes/Links*
Monticello Mill Superfund Site, Monticello, UT	Superfund Former DOE Processing Facility	A cover system was constructed to contain radioactive material removed from the site. The cover design mimics and enhances the natural ground water balance and uses a capillary barrier. Native vegetation was planted to maximize evapotranspiration.	The native vegetation chosen was designed to emulate the structure, function, diversity, and dynamics of native plant communities in the area.	Not specified	Not specified	Mark Aguilar EPA Region 8 1595 Wynkoop Street Mail Code: 8EPR-F Denver, CO 80202-1129 303-312-6251 aguilar.mark@epa.gov	http://www.clu-in.org/PRODUCTS/NEWSLTRS/ttrend/view.cfm?issue=tt0500.htm
Rocky Flats Plant, Golden, CO	Superfund Former DOE Weapons Facility	At one time the site stored more than 14 tons of plutonium. All special nuclear materials were packaged and shipped to licensed repositories. Over 800 structures were cleaned up, as necessary, and removed. 690 tanks were decontaminated and removed, and onsite landfills were covered. Three	Part of the site that has been remediated has been transferred from DOE to DOI and the USFWS to manage as a National Wildlife Refuge.	Not specified	Not specified	Mark Aguilar EPA Region 8 1595 Wynkoop Street Mail Code: 8EPR-F Denver, CO 80202-1129 303-312-6251 aguilar.mark@epa.gov	http://www.epa.gov/region8/superfund/co/rkyflatsplant/index.html

(Continued)

Property Name and Location	Property Type	Cleanup Type	Revitalization/Reuse Component	Problems/Issues	Solutions	Point of Contact	Notes/Links*
		contaminated ground water plume barriers and passive treatment systems were installed. Finally, wastes and contaminated soils were removed and shipped to permitted facilities.					
Rocky Mountain Arsenal, Commerce City, CO	Superfund Army-Lead Remedial Action Ammunition Plant	P&T systems were installed to remediate ground water contaminated with wastes from production of chemical warfare agents, industrial and agricultural chemicals, and pesticides.	Congress passed the Rocky Mountain Arsenal National Wildlife Refuge Act, requiring the site to become part of the national wildlife refuge system once cleanup is complete.	Not specified	Not specified	Greg Hargreaves, RPM EPA Region 8 1595 Wynkoop Street Mail Code: 8EPR-F Denver, CO 80202-1129 303-312-6661 hargreaves.greg@epa.gov	http://www.rma.army.mil/cleanup/clnfrm.html
Silver Bow Creek and Warm Springs Ponds, Butte, MT	Superfund Mining Site	Remediation included excavating sediment contaminated by copper mining activities and installing a water treatment system.	Extensive wetlands are now home to a variety of wildlife. Nesting platforms were built to protect birds. The wetlands are also used for	Not specified	Not specified	Ron Bertram, RPM EPA Region 8 1595 Wynkoop Street Mail Code: 8EPR-F Denver, CO 80202-1129 406-441-1150 bertram.ron@epa.gov	http://cfpub.epa.gov/supercpad/cursites/csitinfo.cfm?id=0800416

(Continued)

Property Name and Location	Property Type	Cleanup Type	Revitalization/Reuse Component	Problems/Issues	Solutions	Point of Contact	Notes/Links*
			recreation such as fishing, hiking, and biking.				
Summitville Mine, CO	Superfund Mining Site	Gold mining released cyanide and acidic mine water to the Alamosa River. Cleanup activities include permanently stabilizing the site and reversing the effects of mining on the river.	The Alamosa River and tributaries flow through wetlands, forested and agricultural land, and into the Terrace Reservoir, which supplies irrigation water to livestock and farms. The site has been revegetated with grasses that promote the recolonization of native plants. The river, which was void of life because of contamination, now supports some types of aquatic life.	Not specified	Not specified	Victor Ketellapper, RPM EPA Region 8 1595 Wynkoop Street Mail Code: 8EPR-F Denver, CO 80202-1129 303-312-6578 ketellapper.victor@epa.gov	http://cfpub.cpa.gov/supercpad/cursites/csitinfo.cfm?id=0801194
REGION 9							
Atlas Asbestos Mine, Fresno County, CA	Superfund Mining Site	The remedy included the removal of contaminated material, stabilization	The site is a wildlife sanctuary and a popular recreational area for hikers,	At the Atlas Mine Area, the road to the Rover	Alternate access roads to the Rover Pit/Channel	Anna Lynn Suer EPA Region 9 75 Hawthorne Street Mail Code: WTR-2	http://www.epa.gov/superfund/sites/fiveyear/

(Continued)

Property Name and Location	Property Type	Cleanup Type	Revitalization/Reuse Component	Problems/Issues	Solutions	Point of Contact	Notes/Links*
		of erosion- prone areas, and structural improvements to clean up the asbestos contaminated soil and water.	campers, and hunters.	Pit/Channel A is likely to fail sometime in the future due to an active landslide.In addition, the road to Pond A may also fail in the future due to erosion.	A and to Pond A will be identified prior to failure of the existing roads.	San Francisco, CA 94105 415-972-3148 suer.lynn@epa.gov	f2006090001 092 .pdf
A West Coast Refinery, Location not provided	EPA Research Technology Development Forum Site Refinery Effluent Treatment System	A phytoremediation demonstration was conducted at the site, which was contaminated with hydrocarbons. The remediation also included enhancing and planting wetlands, and installing a vegetation cap.	The site includes a clean stormwater holding basin. Natural vegetation was planted over the 90-acre vegetation cap.	Selenium was identified on site and in bird eggs, which can be harmful to the wildlife, especially bird embryos.	The site was turned into a treatment zone and habitat zone. Birds were discouraged from the treatment zone where selenium was to be removed. After testing, selenium was found to be greatly reduced in bird eggs.	Kim Beman Chevron 6001 Bollinger Canyon Road San Ramon, CA 94583, KBGS@chevron.com	http://www. wildlifehc.org/ eweb editpro/items/ O57F3055.pdf

(Continued)

Property Name and Location	Property Type	Cleanup Type	Revitalization/Reuse Component	Problems/Issues	Solutions	Point of Contact	Notes/Links*
Alameda Naval Air Station, Alameda, CA	Superfund Landfill, Lagoon	Remediation included using dredged sediment from the lagoon as part of a landfill cap for parts of the site that were contaminated with PCBs, heavy metals, and PAHs.	A golf course is being planned in the landfill area, and a marina will be constructed in the lagoon area.	Not specified	Not specified	Anna Marie Cook EPA Region 9 75 Hawthorne Street Mail Code: SFD-8-3 San Francisco, CA 94105 415-972-3029 cook.anna-marie@epa.gov	http://www.epa.gov/oerrpage/superfund/programs/recycle/pilot/facts/r9 38.htm
REGION 10							
American Crossarm & Conduit Co., Chehalis, WA	Superfund Wood Treatment Facility	Remediation activities include removing contaminated site material, disposing of the site facilities, removing lagoon sediment, and excavating soil. The contaminants of concern are carcinogenic polyaromatic hydrocarbons, PCP, and dioxin/furans.	Wetlands restoration.	Not specified	Not specified	Anne McCauley EPA Region 10 1200 Sixth Avenue Mail Code: ECL-1 13 Seattle, WA 98101 206-553-4689 mccauley.anne@epa.gov	http://www.epa.gov/superfund/sites/fiveyear/f04-1 0004.pdf
Commencement Bay, Tacoma, WA	Superfund Industrial Activities	Industrial activities resulting in hazardous waste contamination of the waterways	In addition to navigational improvements to the port, nine acres of	Not specified	Not specified	Chris Bellovary EPA Region 10 1200 Sixth Avenue Mail Code: ECL-1 11 Seattle, WA 98101	http://cfpub.epa.gov/supercpa

(Continued)

Property Name and Location	Property Type	Cleanup Type	Revitalization/Reuse Component	Problems/Issues	Solutions	Point of Contact	Notes/Links*
		within Commencement	wetlands were restored as a result of the cleanup. EPA also worked with Washington Department of Environment to create seven acres of essential mud flats habitat where fish, birds, wildlife, and plant species thrive.			206-553-2723 bellovary.chris@epa.gov	d/cursites/csitinfo.cfm?id=1000981
Harmony Mine and Mill, Baker, ID	Superfund Mining Site	A diversion ditch was created and pipes laid to divert Withington Creek from tailings piles. After they were dry, 10,000 cubic yards of tailings were excavated and hauled to a repository location. A sedimentation pond was also constructed below the tailings pile to catch any runoff that occurred. Tailings were then capped	Where the tailings were removed, the area was graded, a stable creek bed with the ability to withstand large debris flow was constructed, and disturbed areas were seeded. Withington Creek is a designated cold water community and salmonid spawning habitat for the.	Not specified	Not specified	Greg Weigel EPA Region 10, Idaho Operations Office 1435 North Orchard Street Boise, ID 83706 208-378-5773 weigel.greg@epa.gov	http://epaosc.net/siteprofile.asp?site id=10BN

(Continued)

Property Name and Location	Property Type	Cleanup Type	Revitalization/Reuse Component	Problems/Issues	Solutions	Point of Contact	Notes/Links*
		with a 2-foot layer of compacted rock followed by a one-foot layer of uncompacted rock.	endangered chinook salmon				
Hoquarton Natural Interpretive Trail, Tillamook, OR	Brownfields Lumber Mill	Using an EPA Revolving Loan Fund, contaminated soil was excavated and treated.	The former lumber mill was transformed into a recreational and educational greenspace. Volunteers removed weeds and invasive plants, disposed of over two tons of trash, and planted over 2,000 native plants in riparian areas. A trail was also installed to provide walking and bird watching opportunities.	It was unclear how long-term maintenance of the park would be achieved.	Long-term maintenance of the park was supported by school groups and other volunteers.	Mike Slater EPA Region 10 805 SW Broadway Mail Code: OOO Portland, OR 97205 503-326-5872 slater.mike@epa.gov	http://www.landcurrent.com/contemporary/landscapedesign.php?in=Hoquarton&work=public
Old Jensen Texaco Station, Rosalia, WA	OUST Abandoned Gas Station	Through the USTFields Pilot Program, this abandoned gas station site was remediated by removing five	Stakeholders plan to convert the former gas station site into a visitor and community center with green	Additional contamination could not be removed without destroying the historic building	Not specified	Wildlife Habitat Council 8737 Colesville Road, Suite 800 Silver Spring, MD 20910 301-588-8994 whc@wildlifehc.org	http://www.wildlifehc.org/ewebeditpro/items/O57F7008.pdf

(Continued)

Property Name and Location	Property Type	Cleanup Type	Revitalization/Reuse Component	Problems/Issues	Solutions	Point of Contact	Notes/Links*
		USTs and contaminated soil to make the site ready for future reuse. Contaminated soil treated and disposed of off-site. Additional contamination is being addressed through ground water monitoring and possible MNA.	infrastructure. They plan to incorporate native plant communities that are part of the the distinctive Palouse ecosystem, including grasslands, scrub thickets, ridges, and slope communities. The community center could be used to educate visitors about the unique geology and ecology of the region.	this project was intended to restore. *In situ* treatment options have been considered but will not be pursued until additional ground water data is evaluated. MNA of the remaining contamination may prove to be an adequate and appropriate cleanup alternative.			

* Links valid at time of publication.

APPENDIX B. ADDITIONAL ECOLOGICAL REVITALIZATION RESOURCES

Section 1. Introduction

Interstate Technology & Regulatory Council (ITRC): www.itrcweb.org

Land Revitalization Initiative: *www.epa.gov/oswer/landrevitalization/ basicinformation.htm*

U.S. Environmental Protection Agency (EPA) Hazardous Waste Cleanup Information (CLU-IN). Tools for Ecological Land Reuse: www.cluin.org/ecotools

EPA One Cleanup Program Initiative: www.epa.gov/oswer/ onecleanupprogram

Section 2. Ecological Revitalization Under EPA Cleanup Programs

Atlas Tack Superfund Site Information: www.epa.gov/ne/superfund/ sites/atlas

Brownfields Green Infrastructure Fact Sheet: *www.epa.gov/brownfields/ publications/ swdp0408.pdf*

Biological Technical Assistance Groups (BTAG) Regional Web sites:

EPA Region 3: www.epa.gov/reg3hwmd/risk/eco/index.htm

EPA Region 4: www.epa.gov/region4/waste

EPA Region 5: www.epa.gov/region5superfund/ecology

EPA Region 8: www.epa.gov/region8/r8risk/eco.html

Cross Program Revitalization Guidance:

www.epa.gov/superfund/programs/recycle/pdf/cprm guidance.pdf

Emergency Response Team: www.ert.org

EPA CLU-IN Publications Search Web site: www.clu-in.org/pub1.cfm

EPA CLU-IN Tools for Ecological Land Reuse: www.cluin.org/ecotools

EPA Guidelines for Ecological Risk Assessment: *http://cfpub.epa.gov/ncea/cfm/ recordisplay.cfm?deid=12460*

EPA Land Revitalization Web site:

www.epa.gov/landrevitalization/index.htm

EPA Office of Superfund Remediation and Technology Innovation: www.epa.gov/tio

EPA Region 3—Hazardous Waste Cleanup Sites Land Use & Reuse Assessment, Data Results: www.epa.gov/region03/revitalization/R3_land_use_final/data_results.pdf

EPA Office of Solid Waste and Emergency Response (OSWER). 1991. ECO Update—The Role of Biological Technical Assistance Groups (BTAG) in Ecological Assessment. Publication number 9345.0- 051. September. www.epa.gov/oswer/riskassessment/ecoup/ pdf/v1no1.pdf

EPA OSWER. 2008. Green Remediation: Incorporating Sustainable Environmental Practices into Remediation of Contaminates Sites. www.clu-in.org/download/remed/Green-Remediation-Primer.pdf

Federal Facilities Restoration and Reuse Office (FFRRO) Web site: www.epa.gov/ fedfac/about_ffrro.htm

Interim Guidance for OSWER Cross-Program Revitalization Measures: www.epa.gov/landrevitalization/docs/cprmguidance-10-20-06covermemo.pdf

Local native plant societies: www.michbotclub.org/links/native plant society.htm

National Oceanic and Atmospheric Administration (NOAA): http://response.restoration.noaa.gov

Superfund Sitewide Ready-for-Reuse Performance Measure: www.epa.gov/superfund/programs/recycle/pdf/sitewide_a.pdf

Underground Storage Tank (UST) Brownfields Cleanups: www.nemw.org/petroleum%20issue%20opportunity%20brief.pdf

U.S. Department of Agriculture (USDA), Natural Resources Conservation Service (NRCS): www.nrcs.usda.gov

Wildlife Habitat Council (WHC) Leaking Underground Storage Tank (LUST) Cleanups Web site: www.wildlifehc.org/brownfield restoration/lust pilots.cfm

Section 3. Technical Considerations for Ecological Revitalization

EPA CLU-IN. The Use of Soil Amendments for Remediation, Revitalization, and Reuse: www.clu-in.org/download/remed/epa-542-r-07-013.pdf

EPA Tech Trends. Fort Wainwright: www.clu-in.org/PRODUCTS/ NEWSLTRS/ttrend/view.cfm?issue=tt0500.htm.

Section 4. Wetlands Cleanup and Restoration

EPA, Office of Water, Office of Wetlands, Oceans, and Watersheds: www.epa.gov/OWOW/wetlands

EPA OSWER. Considering Wetlands at Comprehensive Environmental Response, Compensation, and Liability Act (CERCLA) Sites (EPA 540/R-94/019, 1994): www.epa.gov/superfund/policy/remedy/pdfs/540r-94019-s.pdf

EPA OSWER. Environmental Fact Sheet: Controlling the Impacts of Remediation Activities in or Around Wetlands (EPA 530-F-93-020).

Society of Wetland Scientists (SWS), Wetlands Journal: www.sws.org/wetlands

U.S. Department of Interior (DOI), U.S. Fish and Wildlife Service. National Wetlands Inventory: www.nwi.fws.gov

U.S. Geological Survey (USGS), National Wetlands Research Center: www.nwrc.gov

Wetlands Research Program and Wetlands Research Technology Center: http: / /el.erdc.usace.army.mil/wetlands

Wetland Science Institute, Natural Resources Conservation Service, U.S. Department of Agriculture: www.wli.nrcs.usda.gov

Section 5. Stream Cleanup and Restoration

EPA Office of Water. River Corridor and Wetland Restoration Web site: www.epa.gov/owow/wetlands

EPA Office of Water and OSWER. Integrating Water and Waste Programs to Restore Watersheds: www.epa.gov/superfund/resources

EPA OSWER. Contaminated Sediment Remediation Guidance: www.epa.gov/superfund/health/conmedia/sediment

Federal Interagency Stream Corridor Restoration Guide: www.nrcs.usda.gov/ technical/stream_restoration/newgra.html

University of Nebraska-Lincoln: www.ianr.unl.edu/pubs/Soil/g1307.htm

Section 6. Terrestrial Ecosystems Cleanup and Revitalization

Clemants, Stephen. 2002. Is Biodiversity Sustainable in the New York Metropolitan Area? University Seminar on Legal, Social, and Economic Environmental Issues, Columbia University, December 2002.

EPA OSWER. 2008. Green Remediation: Incorporating Sustainable Environmental Practices into Remediation of Contaminates Sites. www.clu-in.org/download/remed/Green-Remediation-Primer.pdf

Handel, Steven N., G.R. Robinson, WFJ Parsons, and J.H. Mattei. 1997. Restoration of Woody Plants to Capped Landfills: Root Dynamics in an Engineered Soil, Restoration Ecology, 5:178-186.

North Carolina Cooperative Extension Service: www.ces.ncsu.edu/depts/hort/hil/hil-645.html Plant Conservation Alliance: www.nps.gov/plants

Robinson, G.R. and S.N. Handel. 1993. Forest Restoration on a Closed Landfill: Rapid Addition of New Species by Bird Dispersion, Conservation Biology, 7: 271-278.

Society for Ecological Restoration. Ecological Restoration Reading Resources: www.ser.org/reading resources.asp

USDA, NRCS. Plant Materials Program: http://plant-materials.nrcs.usda.gov

USDA, NRCS. PLANTS Database: http://plants

Weed Science Society of America: www.wssa.net

Section 7. Long-Term Stewardship Considerations

EPA. Superfund – Operation and Maintenance Web site: http://epa.gov/superfund/cleanup/postconstruction/operate.htm

EPA OSWER. 2005. Long Term Stewardship Task Force Report and the Development of Implementation Options for the Task Force Recommendations. www.epa.gov/LANDREVITALIZATION/docs/ltsreport-sept2005.pdf.

Institutional Controls: A Site Manager's Guide to Identifying, Evaluating, and Selecting Institutional Controls at Superfund and RCRA Corrective Action Cleanups, available at http://epa.gov/superfund/policy/ ic/guide/guide.pdf

APPENDIX C. ACRONYMS

ACRES	Assessment, Cleanup, and Redevelopment Exchange System	FFRRO	Federal Facilities Restoration and Reuse Office
AOC	Area of Concern	FS	Feasibility Study
BMP	Best Management Practices	FY	Fiscal Year
BP	British Petroleum	GPRA	Government Performance and Results Act
BRAC	Base Realignment and Closure	HE EI	Human Exposures Under Control Environmental Indicator
BTAG	Biological Technical Assistance Group	HMX	High Melting Explosive (or Cyclotetramethylenetetranitramine)
BTEX	Benzene, Toluene, Ethylbenzene, and Xylenes	IC	Institutional Control
CERCLA	Comprehensive Environmental Response, Compensation, and Liability Act	IEPA	Illinois Environmental Protection Agency
CERCLIS	Comprehensive Environmental Response, Compensation, and Liability Information System	ITRC	Interstate Technology & Regulatory Council
CIC	Community Involvement Coordinator	JOAAP	Joliet Army Ammunition Plant Design
CLU-IN	Hazardous Waste Clean-up Information	LEED	Leadership in Energy and Environment
CPRM	Cross-Program Revitalization Measure	LUST	Leaking Underground Storage Tank
DARRP	Damage Assessment, Remediation and Restoration Program	MCL	Maximum Contaminant Level
DEQ	Department of Environmental Quality	MDEQ	Michigan Department of Environmental Quality
DNT	Dinitrotoluene	MNA	Monitored Natural Attenuation
DoD	U.S. Department of Defense	NOAA	National Oceanic and Atmospheric Administration
DOE	U.S. Department of Energy	NPL	National Priorities List
DOI	U.S. Department of Interior	NRC	National Research Council
EO	Executive Order	NRCS	Natural Resources Conservation Service
EOD	Explosives Ordnance Disposal	NRDA	Natural Resource Damage Assessment
EPA	U.S. Environmental Protection Agency	O&M	Operation and Maintenance
ER3	Environmentally Responsible Redevelopment and Reuse	OBLR	Office of Brownfields and Land Revitalization
ERA	Ecological Risk Assessment	OPEI	Office of Policy, Economics, and Innovation
FFEO	Federal Facilities Enforcement Office	ORCR	Office of Resource Conservation and Recovery
FFLC	Federal Facilities Leadership Council	OSC	On-Scene Coordinator
OSWER	Office of Solid Waste and Emergency Response	OSRTI	Office of Superfund Remediation and Technology Innovation

Appendix C (Continued)

OU	Operable Unit	PDF	Portable Document Format
OUST	Office of Underground Storage Tanks	PFP	Protective For People
P&T	Pump and Treat	RAU	Ready for Anticipated Use
PAH	Polycyclic Aromatic Hydrocarbon	RCRA	Resource Conservation and Recovery Act
PCA	Plant Conservation Alliance	RDX	Royal Demolition Explosive (or Cyclotrimethyl enetrinitramine)
PCB	Polychlorinated Biphenyl	RMA	Rocky Mountain Arsenal
PCE	Perchloroethylene (or Tetrachloroethene)	ROD	Record of Decision
RI/FS	Remedial Investigation/Feasibility Study	RPM	Remedial Project Manager
RTU	Return To Use	SRI	Superfund Redevelopment Initiative
SVOC	Semi-Volatile Organic Compound	SWS	Society of Wetland Scientists
TAB	Technical Assistance to Brownfields	TCE	Trichloroethylene
TNT	Trinitrotoluene	TPM	Technical Performance Measure
USACE	U.S. Army Corps of Engineers	USDA	U.S. Department of Agriculture
USFWS	U.S. Fish and Wildlife Service	USGS	U.S. Geological Survey
UST	Underground Storage Tank	UXO	Unexploded Ordnance
VOC	Volatile Organic Compound	WHC	Wildlife Habitat Council

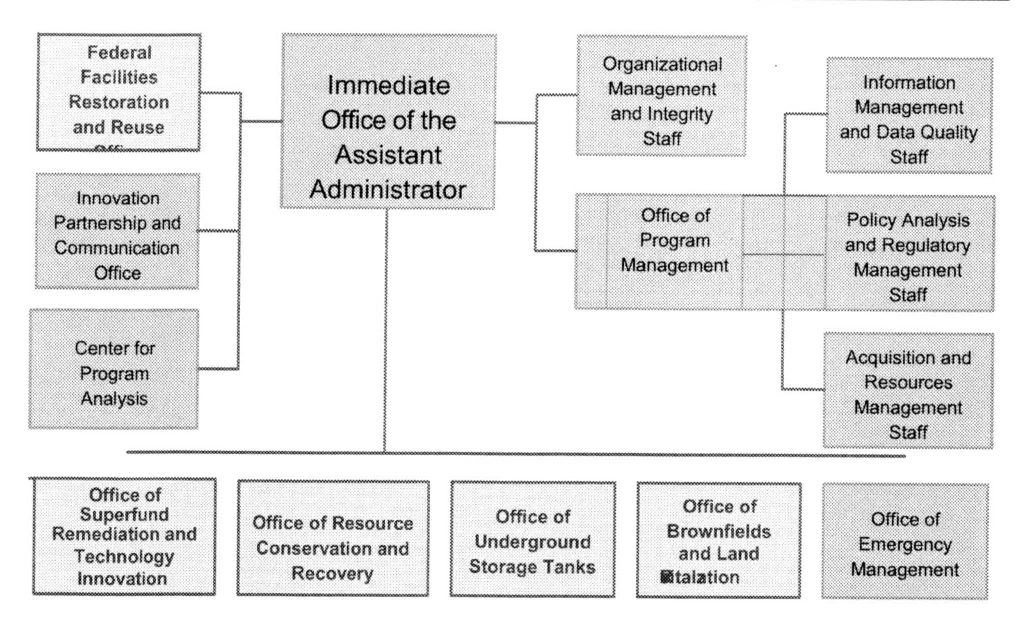

Note: Highlighted EPA offices contributed to the development of this document.

End Notes

[1] "Real property" is a legal term indicating a property consisting of lands and of all appurtenances to lands, as buildings, crops, or mineral rights (distinguished from personal property).

In: Ecological Revitalization and Green Remediation ... ISBN: 978-1-61122-520-4
Editor: Eric S. López

Chapter 2

GREEN REMEDIATION: INCORPORATING SUSTAINABLE ENVIRONMENTAL PRACTICES INTO REMEDIATION OF CONTAMINATED SITES

United States Environmental Protection Agency

ACKNOWLEDGMENTS

The *Green Remediation: Incorporating Sustainable Environmental Practices into Remediation of Contaminated Sites* technology primer was developed by the United States Environmental Protection Agency (U.S. EPA) Office of Superfund Remediation and Technology Innovation (OSRTI). The document was prepared in cooperation with EPA's Brownfields and Land Revitalization Technology Support Center (BTSC) and funded by OSRTI under Contract No. 68-W-03-038 to Environmental Management Support, Inc. The authors gratefully acknowledge the insightful comments and assistance of reviewers within EPA and other federal and state environmental agencies.

ACRONYMS AND ABBREVIATIONS

ARAR applicable or relevant and appropriate requirement
BMP best management practice
CERCLA Comprehensive Environmental Response, Compensation, and Liability Act of 1 980, as amended
CH_4 methane
CO_2 carbon dioxide
CSP concentrating solar power
DOD U.S. Department of Defense
DOE U.S. Department of Energy

EERE	U.S. DOE Office of Energy Efficiency and Renewable Energy
EPA	U.S. Environmental Protection Agency
ET	evapotranspiration
FY	fiscal year
GHG	greenhouse gas
IDW	investigation derived waste
kW	kilowatt
kWh	kilowatt-hour
LEED	Leadership in Energy and Environmental Design
LFG	landfill gas
LID	low impact development
MNA	monitored natural attenuation
mph	miles per hour
MW	megawatt
N_2O	nitrous oxide
NCP	National Oil and Hazardous Substances Pollution Contingency Plan
NPL	National Priorities List
NREL	U.S. DOE National Renewable Energy Laboratory
O&M	operation and maintenance
OSRTI	U.S. EPA Office of Superfund Remediation and Technology Innovation
OSWER	U.S. EPA Office of Solid Waste and Emergency Response
P&T	pump-and-treat
PRB	permeable reactive barrier
PV	photovoltaic
RCRA	Resource Conservation and Recovery Act of 1976, as amended
ROD	record of decision
RSE	remedial system evaluation
SVE	soil vapor extraction
UST	underground storage tank
UV	ultraviolet
VOC	volatile organic compound
WTE	waste-to-energy

SECTION 1: INTRODUCTION

As part of its mission to protect human health and the environment, the U.S. Environmental Protection Agency (EPA or "the Agency") is dedicated to developing and promoting innovative cleanup strategies that restore contaminated sites to productive use, reduce associated costs, and promote environmental stewardship. EPA strives for cleanup programs that use natural resources and energy efficiently, reduce negative impacts on the environment, minimize or eliminate pollution at its source, and reduce waste to the greatest extent possible in accordance with the Agency's strategic plan for compliance and environmental stewardship (U.S. EPA Office of the Chief Financial Officer, 2006). The practice of "green remediation" uses these strategies to consider all environmental effects of remedy implementation for contaminated sites and incorporates options to maximize the net environmental benefit of cleanup actions.

EPA's regulatory programs and initiatives actively support site remediation and revitalization that result in beneficial reuse such as commercial operations, industrial facilities, housing, greenspace, and renewable energy development. The Agency has begun examining opportunities to integrate sustainable practices into the decision-making processes and implementation strategies that carry forward to reuse strategies. In doing so, EPA recognizes that incorporation of sustainability principles can help increase the environmental, economic, and social benefits of cleanup.

Green Remediation: The practice of considering all environmental effects of remedy implementation and incorporating options to maximize net environmental benefit of cleanup actions.

Green remediation reduces the demand placed on the environment during cleanup actions, otherwise known as the "footprint" of remediation, and avoids the potential for collateral environmental damage. The potential footprint encompasses impacts long known to affect environmental media:

- Air pollution caused by toxic or priority pollutants such as particulate matter and lead,
- Water cycle imbalance within local and regional hydrologic regimes,
- Soil erosion and nutrient depletion as well as subsurface geochemical changes,
- Ecological diversity and population reductions, and
- Emission of carbon dioxide (CO_2), nitrous oxide (N_2O), methane (CH_4), and other greenhouse gases contributing to climate change.

Opportunities to increase sustainability exist throughout the investigation, design, construction, operation, and monitoring phases of site remediation regardless of the selected cleanup remedy. As cleanup technologies continue to advance and incentives evolve, green remediation strategies offer significant potential for increasing the net benefit of cleanup, saving project costs, and expanding the universe of long-term property use or reuse options without compromising cleanup goals.

Purpose of Primer

This primer outlines the principles of green remediation and describes opportunities to reduce the footprint of cleanup activities throughout the life of a project. Best management practices (BMPs) outlined in this document help decision-makers, communities, and other stakeholders (such as project managers, field staff, and engineering contractors) identify new strategies in terms of sustainability. These strategies complement rather than replace the process used to select primary remedies that best meet site-specific cleanup goals. The primer identifies the range of alternatives available to improve sustainability of cleanup activities and helps decision-makers balance the alternatives within existing regulatory frameworks. To date, EPA's sustainability initiatives have addressed a broader scope or focused on selected elements of green remediation such as clean energy.

The primer strives to cross educate remediation and reuse decision-makers and other stakeholders about green remediation using a "whole-site" approach that reflects reuse goals. Greater awareness of the opportunities helps remediation decision-makers address the role of cleanup in community revitalization, and helps revitalization project managers maintain an active voice during all stages of remediation decision-making. To maximize sustainability, cleanup and reuse options are considered early in the planning process, enabling BMPs during remediation to carry forward (Figure 1).

Best practices can be incorporated into all phases of remediation, including site investigation, remedy construction, operation of treatment systems, monitoring of treatment processes and progress, and site close-out. Site-specific green remediation strategies can be documented in service or vendor contracts as well as project materials such as site management plans.

To help navigate the range of green remediation opportunities, this primer provides tools for daily operations and introductory information on the use of renewable energy resources. Profiles of site- specific implementation of green remediation strategies are provided throughout the document to help federal and state agencies, local communities, and other stakeholders learn from collective experiences and successes. As new information becomes available, additional profiles will be available online on EPA's Green Remediation web site (http://www.cluin.org/greenremediation). The document also describes the rapidly expanding selection of incentives for strategy implementation and provides a list of additional resources [bracketed number resources] in addition to direct (parenthetical) references.

Overview of Green Remediation

Strategies for green remediation rely on sustainable development whereby environmental protection does not preclude economic development, and economic development is ecologically viable today and in the long run. The Agency has compiled information from a range of EPA programs supporting sustainability along the categories of the built environment; water, ecosystems and agriculture; energy and environment; and materials and toxics. [General Resource 1, Section 8] Many programs, tools, and incentives are available to help governments, businesses, communities, and individuals serve as good environmental stewards, make sustainable choices, and effectively manage resources.

Sustainable development meets the need of the present without compromising the need of future generations, while minimizing overall burdens to society.

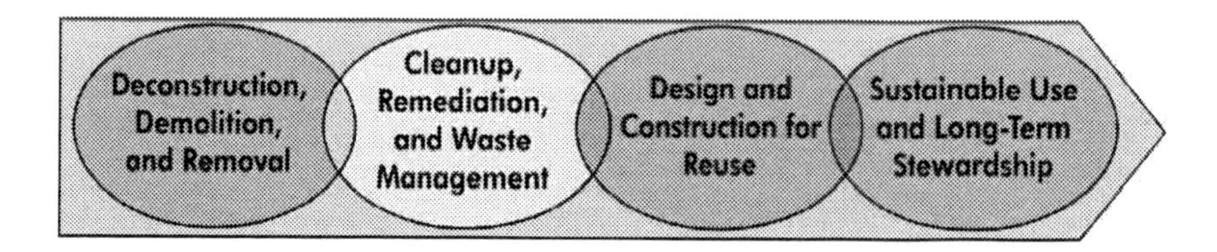

Figure 1. BMPs of green remediation may be used throughout the stages of land revitalization, as a contaminated site progresses toward sustainable reuse or new use

Use of green remediation BMPs helps to accelerate the pace of environmental protection in accordance with the Agency's strategic plan for improving environmental performance of business sectors. Green remediation builds on environmentally conscious practices already used across business and public sectors, as fostered by the Agency's Sectors Program, and promotes incorporation of state-of-the-art methods for:

- Conserving water,
- Improving water quality,
- Increasing energy efficiency,
- Managing and minimizing toxics,
- Managing and minimizing waste, and
- Reducing emission of criteria air pollutants and greenhouse gases (GHGs) (U.S. EPA National Center for Environmental Innovation, 2006).

Increasing concerns regarding climate change have prompted major efforts across the globe to reduce GHG emissions caused by activities such as fossil fuel consumption. [2] The Agency's current strategic plan calls for significant reductions in GHG emissions as well as increases in energy efficiency as required by federal mandates such as Executive Order 13423: *Strengthening Federal Environmental, Energy, and Transportation Management* (Executive Order 13423, 2007). [3, 4] Accordingly, one category of EPA's evolving practices for green remediation places greater emphasis on approaches that reduce energy consumption and GHG emissions:

- Designing treatment systems with optimum efficiency and modifying as needed,
- Using renewable resources such as wind and solar energy to meet power demands of energy-intensive treatment systems or auxiliary equipment,
- Using alternate fuels to operate machinery and routine vehicles,
- Generating electricity from byproducts such as methane gas or secondary materials, and
- Participating in power generation or purchasing partnerships offering electricity from renewable resources.

BMPs of green remediation help balance key elements of sustainability:

- Resource conservation measured by "water intensity," the amount of water necessary to remove one pound of contaminant, or by "soil intensity," the amount of soil displaced or disturbed to remove one pound of contaminant,
- "Material intensity" measured by the amount of raw materials extracted, processed, or disposed of for each pound of contaminant treated, and
- Energy efficiency measured by the amount of energy needed to remove one pound of contaminant.

Green remediation strategies also reflect increased recognition of the need to preserve the earth's natural hydrologic cycle. Best management of remediation activities includes water conservation measures, stormwater runoff controls, and recycling of treatment process water.

Techniques for maintaining water balance are based on requirements of federal and state ground water protection and management programs and on recent climate-change findings by government agencies and organizations such as the U.S. Department of Agriculture, U.S. Geological Survey, and National Ground Water Association. [5] The strategies build on ground water and surface water management requirements under the Clean Water Act and Safe Drinking Water Act as well as water conservation goals set by Executive Order 13423.

Universe of Sites

Green remediation promotes adoption of sustainable strategies at every site requiring environmental cleanup, whether conducted under federal, state, or local cleanup programs or by private parties. Past spills, leaks, and improper management or disposal of hazardous materials and wastes have resulted in contaminated land, water, and/or air at hundreds of thousands of sites across the country. EPA and its state, tribal, and territorial partners have developed a number of programs to investigate and remediate these sites.

Most federal cleanup programs are conducted under statutory authority of the Resource Conservation and Recovery Act (RCRA) of 1976, as amended by the Hazardous and Solid Waste Amendments of 1 984; Comprehensive Environmental Response, Compensation, and Liability Act of 1 980 (CERCLA), as amended by the Superfund Amendments and Reauthorization Act of 1 986; and Small Business Liability Relief and Brownfields Revitalization Act of 2001. Most states maintain parallel statutes providing for voluntary and mandatory cleanup as well as brownfield and reclamation programs. In addition, most states have attained authority to implement federal mandates under the RCRA corrective action and underground storage tank programs.

Remediation activities in the United States may be grouped into seven major cleanup programs or market segments implemented under different federal or state statues. These market segments are described in *Cleaning Up the Nation's Waste Sites: Markets and Technology Trends*, along with estimates of the number of sites under each major cleanup program (U.S. EPA/OSWER, 2004). Principles and BMPs of green remediation can be applied at sites in each of the market segments, although administrative, institutional, and remedy-selection decision criteria may vary across programs. Based on this report and other summary data, EPA estimates the approximate number of sites requiring remediation under each of the major cleanup programs.

Superfund Sites: As of 2005, nearly 3,000 CERCLA records of decision (RODs) and ROD amendments had been signed. RODs document treatment, containment, and other remedies for contaminated materials at approximately 1 ,300 of the more than 1 ,500 sites historically listed on the National Priorities List (NPL), including those delisted over the years. Superfund cleanups also encompass "removals," which are short-term actions to address immediate threats and emergency responses. Since its inception, the program has undertaken more than 9,400 removal actions.

RCRA Sites: EPA estimates that more than 3,700 regulated hazardous waste treatment, storage, and disposal facilities are expected to need corrective action under the RCRA Corrective Action Program.

Underground Storage Tank Sites: Through September 2007, over 474,000 releases of hazardous substances have been reported at sites with underground storage tanks. Of these, 365,000 cleanups have been completed, leaving approximately 1 09,000 sites with reported releases to be remediated. In recent years, between 7,000 and 9,000 new reports of releases were received annually.

Department of Defense Sites: The U.S. Department of Defense (DOD) estimates that investigations and/or cleanups are planned or underway at nearly 8,000 areas. These areas are located on hundreds of active and inactive installations and formerly used defense sites.

Department of Energy Sites: The U.S. Department of Energy (DOE) has remediated contaminated areas at more than 100 installations and other locations. The Department has identified approximately 4,000 contaminated or potentially contaminated areas on 22 installations and other locations. Most of DOE's remediated areas will require ground water treatment and monitoring or other long-term stewardship efforts.

Other Federal Agency Sites: EPA estimates that there are more than 3,000 contaminated sites, located on 700 federal facilities, potentially requiring remediation. These facilities are distributed among 1 7 federal agencies. Investigations at many of these facilities are not complete. These estimates do not include an estimated 8,000-3 1,000 abandoned mine sites, most of which are located on federal lands.

State, Brownfield, and Private Sites: EPA estimates that during 2006 and 2007 alone cleanups were completed at over 1 8,900 sites, totaling over 250,000 acres, through state and tribal response programs. Institutional controls have also been put in place where required. EPA's investment in brownfields, exceeding 1 .3 billion dollars through 2007, has leveraged more than $10.3 billion in cleanup and redevelopment funding and financed assessment and/or cleanup of more than 4,000 properties.

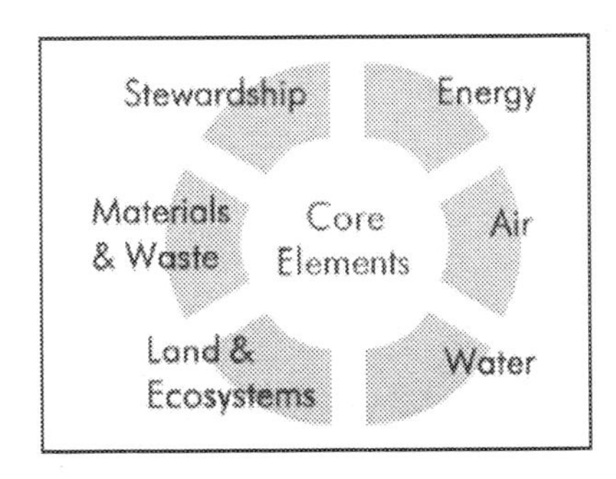

Figure 2. Best management practices of green remediation balance core elements of a cleanup project.

Cleanups across these market segments involve a wide range of pollution sources and site types such as neighborhood dry cleaners and gas stations, former industrial sites in urban areas, metals- contaminated mining sites, and large DOD, DOE, and industrial facilities that are downsized or decommissioned. Cleanup and reuse of these sites will consume significant amounts of energy, considerably impact natural resources, and affect the infrastructures of surrounding communities.

SECTION 2. SUSTAINABILITY OF SITE REMEDIATION

Green remediation focuses on maximizing the net environmental benefit of cleanup, while preserving remedy effectiveness as part of the Agency's primary mission to protect human health and the environment. Site-specific strategies must take into account the unique challenges and characteristics of a site; no single solution exists. At all sites, however, key opportunities for integrating core elements of green remediation can be found when designing and implementing cleanup measures. Regulatory criteria and standards serve as a foundation for building green practices.

Core Elements of Green Rem ediation

Green remediation results in effective cleanups minimizing the environmental and energy footprints of site remediation and revitalization. Sustainable practices emphasize the need to more closely evaluate core elements of a cleanup project; compare the site-specific value of conservation benefits gained by different strategies of green remediation; and weigh the environmental trade-offs of potential strategies. Green remediation addresses six core elements (Figure 2):

Energy requirements of the treatment system

- Consider use of optimized passive-energy technologies (with little or no demand for external utility power) that enable all remediation objectives to be met,
- Look for energy efficient equipment and maintain equipment at peak performance to maximize efficiency,
- Periodically evaluate and optimize energy efficiency of equipment with high energy demands, and
- Consider installing renewable energy systems to replace or offset electricity requirements otherwise met by the utility.

Air emissions

- Minimize use of heavy equipment requiring high volumes of fuel,
- Use cleaner fuels and retrofit diesel engines to operate heavy equipment, when possible,
- Reduce atmospheric release of toxic or priority pollutants (ozone, particulate matter, carbon monoxide, nitrogen dioxide, sulfur dioxide, and lead), and
- Minimize dust export of contaminants.

Water requirements and impacts on water resources

- Minimize fresh water consumption and maximize water reuse during daily operations and treatment processes,
- Reclaim treated water for beneficial use such as irrigation,
- Use native vegetation requiring little or no irrigation, and
- Prevent impacts such as nutrient loading on water quality in nearby water bodies.

Land and ecosystem impacts

- Use minimally invasive in situ technologies,
- Use passive energy technologies such as bioremediation and phytoremediation as primary remedies or "finishing steps," where possible and effective,
- Minimize soil and habitat disturbance,
- Minimize bioavailability of contaminants through adequate contaminant source and plume controls, and
- Reduce noise and lighting disturbance.

Material consumption and waste generation

- Use technologies designed to minimize waste generation,
- Re-use materials whenever possible,
- Recycle materials generated at or removed from the site whenever possible,
- Minimize natural resource extraction and disposal, and
- Use passive sampling devices producing minimal waste, where feasible.

Long-term stewardship actions

- Reduce emission of CO_2, N_2O, CH_4, and other greenhouse gases contributing to climate change,
- Integrate an adaptive management approach into long-term controls for a site,
- Install renewable energy systems to power longterm cleanup and future activities on redeveloped land,
- Use passive sampling devices for long-term monitoring, where feasible, and
- Solicit community involvement to increase public acceptance and awareness of long-term activities and restrictions.

Green Remediation Objectives

- Achieve remedial action goals,
- Support use and reuse of remediated parcels,
- Increase operational efficiencies,
- Reduce total pollutant and waste burdens on the environment,
- Minimize degradation or enhance ecology of the site and other affected areas,
- Reduce air emissions and greenhouse gas production,

- Minimize impacts to water quality and water cycles,
- Conserve natural resources,
- Achieve greater long-term financial return from investments, and
- Increase sustainability of site cleanups.

Green remediation requires close coordination of cleanup and reuse planning. Reuse goals influence the choice of remedial action objectives, cleanup standards, and the cleanup schedule. In turn, those decisions affect the approaches for investigating a site, selecting and designing a remedy, and planning future operation and maintenance of a remedy to ensure its protectiveness.

Site cleanup and reuse can mutually support one another by leveraging infrastructure needs, sharing data, minimizing demolition and earth-moving activities, re-using structures and demolition material, and combining other activities that support timely and cost-effective cleanup and reuse. Early consideration of green remediation opportunities offers the greatest flexibility and likelihood for related practices to be incorporated throughout a project life. While early planning is optimal, green strategies such as engineering optimization can be incorporated at any time during site investigation, remediation, or reuse.

Regulatory Requirements for Cleanup Measures

EPA's green remediation strategies build on goals established by federal statutes and regulatory programs to achieve greater net environmental benefit of a cleanup. Although remedy selection criteria and performance standards vary in accordance with statutory or regulatory authority, goals remain common among the cleanup programs. Section 121 of CERCLA, for example, requires that remedies:

- Protect human health and the environment,
- Attain applicable or relevant and appropriate requirements (ARARs) or provide reasons for not achieving ARARs,
- Are cost effective,
- Utilize permanent solutions, alternative solutions, or resource recovery technologies to the maximum extent possible, and
- Satisfy the preference for treatment that reduces the toxicity, mobility, or volume of the contaminants as opposed to an alternative that provides only for containment. [6]

Pursuant to CERCLA, the National Oil and Hazardous Substances Pollution Contingency Plan (NCP) also identifies nine evaluation criteria to be used in a detailed analysis of cleanup alternatives:

- Overall protection of human health and the environment,
- Compliance with ARARs,
- Long-term effectiveness and permanence,
- Reduction of toxicity, mobility, or volume through treatment,
- Short-term effectiveness,
- Implementability,
- Cost,
- State acceptance, and
- Community acceptance. [7]

Similarly, several evaluation criteria are used under the Agency's RCRA Corrective Action Program to determine the most favorable alternative for corrective measures: long-term reliability and effectiveness; reduction in toxicity, mobility, or volume of wastes; short-term effectiveness; implementability; cost; community acceptance; and state acceptance.

EPA's strategic plan for compliance and environmental stewardship relies on the Agency's cleanup programs to significantly reduce hazardous material use, energy and water consumption, and GHG intensity by 2012. In addition, the Agency's strategy regarding clean air and global climate change calls for collaboration with DOE and organizations to help the United States reduce its GHG intensity from 2002 levels by 1 8% by 2012. These partnerships encourage sound choices regarding energy efficient equipment, policies and practices, and transportation. BMPs of green remediation provide additional tools for making sustainable choices within this statutory, regulatory, and strategic framework.

Expanded Consideration of Energy and Water Resources

Site remediation and revitalization decisions also must comply with more recent federal and state statutes requiring or recommending reductions in energy and water consumption as well as increased use of renewable energy. The Energy Policy Act of 2005, for example, promotes energy conservation nationwide and increases availability of energy supplies. [8] The Act recognizes that energy production and environmental protection are non-exclusive national goals and encourages energy production and demand reduction by promoting new technology, more efficient processes, and greater public awareness (Capital Research, 2005).

A number of policies are in place to ensure that federal activities meet greener objectives. EPA's strategic plan recognizes that implementing provisions of the Energy Policy Act is a major undertaking involving increased partnership with DOE. DOE's Office of Energy Efficiency and Renewable Energy (EERE) reports that the Act's major provisions, as strengthened by Executive Order 1 3423, require federal facilities (sites owned or operated by federal agencies) to:

- Reduce facility energy consumption per square foot (a) 2% each year through the end of 2015 or a total of 20% by the end of fiscal year (FY) 2015 relative to 2003 baseline; and (b) 3% per year through the end of 2015 or a total of 30% by the end of FY 2015 relative to 2003 baseline (including industrial and laboratory facilities),
- Expand use of renewable energy to meet (a) no less than 3% of electricity demands in FY 2007- 2009, 5% in FY 201 0-FY 2012, and 7.5% in 2013 and thereafter; and (b) at least 50% of the renewable energy requirements through new renewable sources,
- Reduce water consumption intensity by 2% each year through the end of FY 2015 or 16% by the end of FY 201 5 (relative to 2007 baseline) beginning in 2008,
- Employ electric metering in federal buildings by 2012,
- Apply sustainable design principles for building performance standards, and
- Install 20,000 solar energy systems by 201 0.

The Energy Independence and Security Act of 2007 sets additional goals regarding energy consumption and associated GHG emissions, including increased use of alternative fuels for vehicles and new standards for energy efficiency in buildings. [9] The Act also promotes accelerated research and development of alternative energy resources (primarily solar, geothermal, and marine energy technologies) and provides grants to develop technologies for large-scale CO2 capture from industrial sources. To date, 24 states plus the District of Columbia have implemented policies for renewable portfolio standards requiring electricity providers to obtain a minimum percentage of their power from renewable energy resources by a certain date. Four additional states have established non-regulatory goals for adopting renewable energy. [10]

Federal agencies such as the EPA, DOD, DOE, U.S. Department of Agriculture, and General Services Administration are working to develop mechanisms for meeting energy and water conservation goals and deadlines across both government and private sectors. Voluntary or required participation in related federal, state, and a growing number of municipal initiatives provides significant opportunities for integrating green practices into site remediation and reuse.

EPA's sustainability strategy encourages "demand-driven" and participatory decision-making using a systematic approach and life-cycle perspective to evaluate chemical, biological, and economic interactions at contaminated sites. Accordingly, EPA is collaborating with public and private partners to establish benchmarks, identify best practices, and develop the models, tools, and metrics needed to reach the goals of green remediation. The Agency also is compiling new information to quantify the net environmental benefit gained by site-specific reductions in fossil fuel consumption and to estimate related contributions in meeting national climate-change goals. On a local level, EPA regions are working with business and community partners to identify site-specific opportunities for demonstrating and applying these practices.

SECTION 3. SITE MANAGEMENT PRACTICES

BMPs of green remediation help ensure that day-to-day operations during all cleanup phases maximize opportunities to preserve and conserve natural resources while achieving the cleanup's mission of protecting human health and the environment. Opportunities to implement the practices are not restricted to cleanups involving media treatment; for example, the practices can apply to removal actions involving primarily institutional controls or short-term soil excavation with offsite disposal. In these cases, the cleanup approach is similar to one used for sustainable and energy efficient construction projects.

Many of the strategies already are used to some degree in site cleanup, although the practices are not necessarily labeled "green." For example, selection of native rather than non-native plants for remedies such as vegetative landfill covers or soil excavation and revegetation significantly reduces the need to consume water for irrigation purposes – one of the key BMPs for water conservation.

Site management plans can specify BMPs for daily operations that meet the goals of green remediation.

Each site management plan can incorporate practices addressing core elements of green remediation with periodic review and update as new opportunities arise. An adaptive approach to site management planning enables early plans, in many cases initiated during emergency removal actions, to be expanded throughout remediation and extended into long-term stewardship controls. Each plan can outline site-specific procedures to:

- Reduce air emissions and energy use,
- Demonstrate water quality preservation and resource conservation,
- Establish near-term improvements to the ecosystem that carry forward into site revitalization, and
- Reduce material consumption and waste generation.

Many of the BMPs and high performance criteria for site management draw on elements of a variety of programs:

- U.S. Green Building Council's Leadership in Energy and Environmental Design (LEED) rating system for new or existing building construction,
- Joint EPA/DOE Energy Star® product ratings, guidelines for energy management in buildings and plants, and general designs for energy efficient commercial buildings,
- EPA's GreenScapes for landscaping approaches that preserve natural resources while preventing waste and pollution, and
- Smart Growth principles helping to reduce urban sprawl. [11-14]

BMPs also stem from new or ongoing federal initiatives to reduce GHG emissions and energy consumption and generally promote green practices and products within market sectors. Examples include joint EPA/DOE recommendations regarding green construction of federal buildings; requirements for General Services Administration procurement of green products and services; and EPA partnership with trade associations of major manufacturing and service sectors such as the construction industry's Associated General Contractors of America. [15-1 7]

Costs for implementing the "extra steps" of green remediation range considerably but can be equal to or below those of conventional cleanup practices, particularly following an initial learning curve. Effective strategies consider site-specific conditions and requirements, long-term investment returns, energy efficiency, and product or service lifecycles. Efficiency improvements under DOE energy- savings performance contracts, for example, are estimated to provide federal net savings of $1.4 billion. The savings result from implementing recommendations of energy service companies under contracts extending up to 25 years (U.S. DOE/EERE, 2007). Site-specific case studies show that BMPs applicable to green remediation can result in immediate and long-term savings:

- Capital costs for a 3-kilowatt (kW) solar system at the Pemaco Superfund site in Maywood, CA, were recovered after one year of operation. Nine months of solar operations provided sufficient electricity to cover one month of operating the site's treatment building, which contains controls for soil heating and ground water pumping and treatment (U.S. EPA/OSWER, 2008(a)).

- Recent engineering optimization of the ground water pumping and treatment system used at the Havertown PCP Site in Havertown, PA, provides a savings of $32,000 each year. Cost reductions are attributed to lower electricity consumption as well as fewer purchases of equipment parts and process chemicals (U.S. EPA/OSWER, 2006).
- Low impact development strategies involving open space preservation and cluster design result in total capital cost savings of 15-80%, according to the majority of 17 case studies conducted by EPA. The savings are generated by reduced costs for site grading and preparation, stormwater infrastructure, site paving, and landscaping (U.S. EPA/Office of Water, 2007).

One example of innovative strategies used to incorporate BMPs common across market sectors is provided by the passive solar biodiesel-storage shed design (Figure 3) developed by Piedmont Biofuels, a North Carolina community cooperative using and encouraging the use of clean, renewable biofuels. Green elements of the design include cob walls comprising sand, clay, and straw to ensure biodiesel storage at interior temperatures remaining above 20° F; a foundation of locally obtained stone mortared with clay; a low-cost galvanized metal roof for heat retention; and a southern overhang to prevent excess solar gain in summer. When needed, portable solar systems can provide electricity to generate additional interior heat. [18]

Incorporating green remediation into cleanup procurement documents is one way to open the door for best practices in the field. In accordance with federal strategies for green acquisition (Executive Order 13423, 2007), purchasing agreements supporting site cleanup and revitalization should give preference to:

- Products with recycled content,
- Biobased products,
- Alternative fuels,
- Hybrid and alternative fuel vehicles,
- Non-ozone depleting substances,
- Renewable energy,
- Water efficient, energy efficient Energy Star® equipment and products with the lowest watt stand-by power, and
- All services that include supply or use of these products.

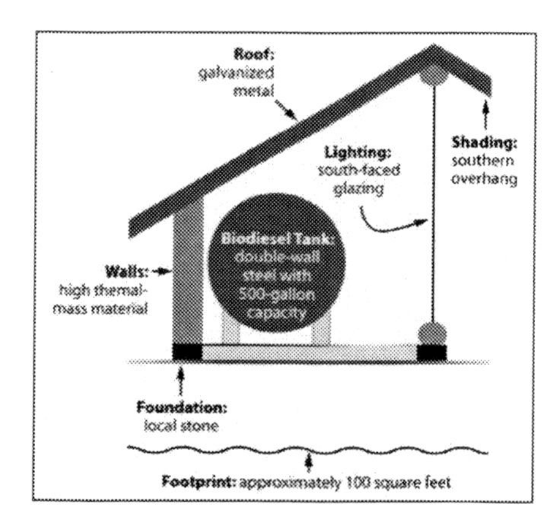

Figure 3. Green construction techniques can be integrated into BMPs for small structures used to store field equipment or to house treatment components such as pump equipment

Site Investigations and Monitoring

Green remediation builds additional sustainability into practices already used for site evaluations and encourages development of novel techniques. Removal actions as well as site assessments and investigations should maximize opportunities for combining field activities in ways that reduce waste generation, conserve energy, and minimize land and ecosystem disturbance. Site investigation and monitoring, including well placement, should consider land reuse plans, local zoning, and maintenance and monitoring of any engineering and institutional controls. BMPs of green remediation help identify sustainable approaches for field work commonly involving subsurface drilling and multimedia data gathering.

At Superfund sites, for example, sampling and analysis plans are required to contain an investigation derived waste (IDW) plan that describes how all ARARS for waste generation and handling will be met, and the best approach for minimizing waste generation, handling, and disposal costs. IDW requirements also apply to projects involving offsite disposal of hazardous waste under other cleanup programs such as RCRA. Typical IDW includes:

- Drilling fluids, cuttings, and purge water from test pits and well installations,
- Purge water, excess soil, and other materials from sample collection,
- Residues such as ash, spent carbon, and well purge water from testing of treatment technologies and aquifer pumping tests,
- Contaminated personal protective equipment, and
- Solutions used to decontaminate non-disposable protective clothing and equipment. [1 9, 20]

Personal protective equipment is usually changed on a daily basis; fewer days in the field result in a smaller quantity of contaminated equipment needing disposal. When cleaning field equipment such as soil and water samplers, drill rods, and augers to prevent contaminant transfer between sample locations, consider using steam and non-phosphate detergent instead of toxic cleaning fluids. Organic solvents and acid solutions should be avoided in decontamination procedures but may be required when addressing free-product contaminants or high concentrations of metals.

Where technically feasible, collection of subsurface soil and ground water samples can rely on direct push drilling rigs rather than conventional rotary rigs. Direct push techniques employ more timesaving tools (particularly for subsurface investigations extending less than 1 00 feet below ground surface), avoid use of drilling fluids, and generate no drill cuttings. Total drilling duration is estimated to be 50-60% shorter for direct push systems. In addition, direct push rigs can be used to collect soil and ground water samples simultaneous to the drilling process. This approach results in reduced IDW volume and field mobilization with related fuel consumption and site disturbance.

Larger push rods now available on the market enable a direct push rig to be used also for placement of monitoring wells with pre-packed screen sizes. This approach provides an alternative to the conventional, energy intensive method involving use of a direct push rig to determine only the location of a long-term monitoring well, and subsequent placement of the well through use of an auger rig. Although some states have not approved wells placed through direct push techniques, this approach to monitoring well installation provides additional fuel and waste savings and significantly reduces the extent of site disturbance. Regardless of drill technique, many rigs operate with diesel engines that can use biodiesel fuel. Site investigations should avoid use of oversized equipment and unnecessary engine idling to maximize fuel conservation.

Geophysical techniques such as ground penetrating radar could be used at some sites to reduce the need for direct measurement of stratigraphic units. Feasibility of using geophysical methods for these purposes depends heavily on site conditions and the nature of contamination. Geophysical surveys result in much smaller environmental footprints than invasive techniques for site investigations, including cone penetrometer test rigs.

BMPs include use of passive sampling techniques for monitoring quality of air, sediment, and ground or surface water over time. In contrast to traditional methods involving infrequent and invasive spot- checking, these methods provide for steady data collection at less cost while generating less waste. Passive techniques for water sampling rely on ambient flow-through in a well without well pumping or purging, avoiding the need for disposal of large volumes of water that require management as hazardous waste. For some contaminants, however, passive devices for obtaining ground water samples are ineffective. [21]

Remote data collection significantly reduces onsite field work and associated labor cost, fuel consumption, and vehicular emissions. For example, water quality data on streams in acid mine drainage areas can be monitored automatically and transmitted to project offices through solar powered telemetry systems. This approach can be used for site investigations as well as site monitoring once treatment is initiated. Renewable energy powered systems with battery backup can be used to operate meteorological stations, air emission sensors, and mobile laboratory equipment. Remote systems also provide quick data access in the event of treatment system breakdown.

Green remediation builds on methods used in the Triad decision-making approach to site cleanup: systematic planning, dynamic work strategies, and real-time measurement systems. The approach advocates onsite testing of samples with submission of fewer samples to offsite laboratories for confirmation. The need for less offsite confirmation saves resources otherwise spent in preserving, packing, and shipping samples overnight to a laboratory. The number of required field samples also can be lowered through comprehensive review of historical information. The Triad approach allows for intelligent decision-making regarding the location and extent of future sampling activities based on the results of completed analytical sampling. This dynamic work strategy significantly minimizes unnecessary analytical sampling. [22]

Air Quality Protection

Green remediation strategies for air quality protection build on requirements or standards under the Clean Air Act, Energy Policy Act, and Energy Independence and Security Act. Cleanup at many sites involves air emissions from treatment processes and often requires use of heavy diesel-fueled machinery such as loaders, trucks, and backhoes to install and sometimes modify cleanup systems (Table 1). BMPs for operation of heavy equipment as well as routine on- or off-road vehicles provide opportunities to reduce emission of GHG and criteria pollutants such as sulfur dioxide. These practices encourage use of new user-friendly tools becoming available from government agencies and industry to help managers estimate and track project emissions.

Contracts for field service can include specifications regarding diesel emissions and air quality controls. Sample language may be drawn from EPA's Clean Construction USA online resource. [23]

Overall efforts should be made to minimize use of heavy equipment and to operate heavy equipment and service vehicles efficiently. Site contracts for service vendors or equipment should give preference to providers able to take advantage of air protection opportunities:

- Retrofitting machinery for diesel-engine emission control and exhaust treatment technologies such as particulate filters and oxidation catalysts,
- Maintaining engines of service vehicles in accordance with manufacturer recommendations involving air filter change, engine timing, and fuel injectors or pumps,
- Refueling with cleaner fuels such as ultra-low sulfur diesel,
- Modifying field operations through combined activity schedules as well as reducing equipment idle, and
- Replacing conventional engines of existing vehicles when feasible, and purchasing new vehicles that are equipped to operate on hybrid systems or alternative fuel and meet the latest engine standards. [24, 25]

Site management plans should specify procedures for minimizing worker and community exposure to emissions, and for minimizing fuel consumption or otherwise securing alternatives to petroleum-based fuel. Plans also should contain specific methods to avoid dust export of contaminants, such as using simple wet-spray techniques, and to control noise from power generation.

Table 1. Mobile sources typically employed during a five-year multi-phase extraction treatment project could consume nearly 30,000 gallons of fuel, equivalent to the amount of carbon annually sequestered by 62 acres of pine or fir forests. [26]

Field Machinery and Vehicles Used for a Typical Multi-Phase Extraction Project	Fuel Consumption (gallons)	CO_2 Emission (pounds)
Site Preparation: One Bobcat with intermittent use of flatbed trailer-truck or dump truck operating for 26 weeks	8,996	199,711
Well Construction: Truck-mounted auger system installing ten 75-foot extraction wells over 30 days	612	13,586
Routine Field Work: Two pickup trucks for site preparation, construction, treatment system monitoring, sampling, and repair over five-year duration	19,760	383,344
Total for Project Life:	**29,368**	**596,641**

Water Quality Protection and Conservation

Best practices for stormwater management limit the disruption of natural water hydrology by reducing impervious cover, increasing onsite infiltration, and reducing or eliminating pollution from stormwater runoff. Green goals used in industry-based programs such as LEED can be applied to cleanup construction; sample targets include:

- Implementing a management plan that results in a 25% decrease in runoff at sites with impervious cover exceeding 50%,
- Capturing 90% of the site's average annual rainfall, and
- Removing 80% of the average annual total load of suspended solids based on pre-construction monitoring reports.

Site "fingerprinting" is an ecology-based planning tool focused on the protection of natural resources during site development.

Site management plans can describe BMPs for reducing and controlling stormwater runoff in manners that mimic the area's natural hydrologic conditions, otherwise known as low impact development (LID). Cleanup at sites undergoing redevelopment could introduce best practices to be used during later stages:

- Conservation designs for minimizing runoff generation through open-space preservation methods such as cluster development, reduced pavement widths, shared transportation access, reduced property setbacks, and site fingerprinting during construction,
- Engineered structures or landscape features helping to capture and infiltrate runoff, such as basins or trenches, porous pavement, disconnected downspouts, and rain gardens or other vegetated treatment systems,
- Storage of captured runoff in rain barrels or cisterns, green (vegetated) roofs, and natural depressions such as landscape islands, and
- Conveyance systems to route excess runoff through and off the site, such as grassed swales or channels, terraces or check dams, and elimination of curbs and gutters. [27]

BMPs reflect maximum efforts to reclaim treated water for beneficial use or re-inject it into an aquifer for storage, rather than discharging to surface water. Where treatment processes result in wastewater discharge to surface water or municipal sewage treatment plants (publicly owned treatment works), green remediation strategies build on criteria of EPA's effluent guidelines. The guidelines rely on industry- proven performance of treatment and control technologies. Best practices for wastewater treatment, including any resulting in pollutant discharge significantly below regulatory thresholds, can be recorded in associated permits for national pollutant discharge elimination systems. [28]

Profile: Old Base Landfill, Former Naval Training Center-Bainbridge, Port Deposit, MD

Cleanup Objectives: Contain an unlined landfill containing nearly 38,000 cubic yards of soil contaminated by waste such as pesticides and asbestos debris

Green Remediation Strategy: Employed BMPs for controlling stormwater runoff and sediment erosion during construction of a landfill cover

- Installed a woven geotextile silt fence down gradient of construction to filter sediment from surface runoff
- Added a "super-silt fence" (woven geotextile with chain-link fence backing) on steep grades surrounding the landfill
- Constructed berms and surface channels to divert stormwater to sediment ponds
- Emplaced erosion control blankets to stabilize slopes and channels until vegetation was established
- Hydroseeded the landfill cover with native seed to foster rapid plant growth

Results:

- Effectively captured sediment at super- silt fence despite heavy rain of Hurricane Floyd
- Avoided damage of infrastructure used in site redevelopment
- Reestablished 100% vegetative cover within one year

Property End Use: Redevelopment for office and light industrial space

BMPs could include estimates of the anticipated demands for potable and non-potable water and substitution of potable with non-potable water whenever possible. One goal might be to replace 50% of the potable water used at a site with non-potable water. Targets can be met by using high efficiency water fixtures, valves, and piping, and by reusing stormwater and greywater for applications such as mechanical systems and custodial operations.

Profile: California Gulch Superfund Site, Leadville, CO

Cleanup Objectives: Address metals- contaminated soil at a former mining site

Green Remediation Strategy: Constructed a recreational trail serving as a cap for contaminated soil

- Conducted a risk-based assessment to confirm trail interception of exposure pathways for waste left in place
- Demonstrated the trail would not harm adjacent wetlands and streams
- Completed a cultural resource inventory and mitigation plan to meet historic preservation requirements
- Consolidated slag-contaminated soil into a platform running along the site's former rail and haul-road corridor

- Covered the soil platform with a six-inch layer of gravel spanning a width of 12 feet with additional three-foot shoulders
- Installed six inches of asphalt above the gravel layer

Results:

- Avoided invasive soil excavation and costly offsite disposal
- Reduced consumption and cost of imported construction material through use of contained waste-in-place
- Increased user safety and remedy integrity through trail restriction to non- motorized use
- Relied on an integrated remediation and reuse plan involving extensive community input, donation of land and construction material by the property owner, and long-term trail and remedy maintenance by Lake County, CO

Property End Use: Recreation

Green remediation practices potentially help cleanups not only meet but exceed water-quality and drinking-water standards set by federal and state agencies. In turn, the benefit of higher water quality can be passed to future site users. Broader strategies for managing a cleanup project's impact on local watershed conditions can complement regional water and waste programs for watershed restoration. [29]

Ecological and Soil Preservation

Green remediation practices provide a whole- site approach that accelerates reuse of degraded land while preserving wildlife habitat and enhancing biodiversity. BMPs can provide novel tools for measuring a site's progress toward meeting both short- and long-term ecological land reuse goals involving:

- Increased wildlife habitat,
- Increased carbon sequestration,
- Reduced wind and water erosion,
- Protection of water resources,
- Establishment of new greenspaces or corridors,
- Increases in surrounding property values, and
- Improved community perception of a site during cleanup. [30]

Site management plans can describe an approach to ecological preservation that considers anticipated reuse as well as the natural conditions prevailing before contamination occurred. BMPs address daily routines that minimize wildlife disturbance, including noise and lights affecting sensitive species. On previously developed or graded sites, goals for habitat restoration might include planting of native vegetation on 50% of the site. Native plants require minimal or no irrigation following establishment and require no maintenance such as mowing or chemical inputs such as fertilizers. Invasive plants or noxious weeds are always prohibited.

Profile: Rhizome Collective Inc. Brownfield Site, Austin, TX

Cleanup Objectives: Clean up illegal dump containing 5,000 cubic yards of debris

Green Remediation Strategy: Constructed a four-foot-thick evapotranspiration cover

- Salvaged wood scraps and concrete for erosion control
- Chipped or shredded wood to create mulch for recreational trails
- Recycled 31.6 tons of metal
- Salvaged concrete for later use as fill for building infrastructure
- Powered equipment through use of bio fuel generators and photovoltaic panels, due to lack of grid electricity
- Extracted 680 tires through use of vegetable oil powered tractor
- Inoculated chainsaws with fungi spore- laden oil to aid in degradation of residual contaminants
- Constructed floating islands of recovered plastic to create habitat for life forms capable of bioremediating residual toxins in an onsite retention pond
- Planted native grasses, wildflowers, and trees

Results:

- Reestablished wildlife habitat for native and endangered species
- Gained community help to restore the property within a single year

Property End Use: Environmental education park

Ecological restoration and preservation at sites anticipated for full or partial reuse as greenspace are best managed through site surveys and careful master planning. BMPs for greenspace could include targets such as confining site disturbance to areas within 1 5 feet of roadways and utility trenches or within 25 feet of pervious areas of paving.

BMPs include development of an erosion and sedimentation control plan for all activities associated with cleanup construction and implementation. Objectives include:

- Preventing loss of soil by stormwater runoff or wind erosion,
- Preventing topsoil compaction, thereby increasing subsurface water infiltration,
- Preventing sediment transport to storm sewers or streams, and
- Preventing dispersion of dust and particulate matter.

Potential strategies for erosion and sedimentation control include stockpiling of topsoil for reuse, temporary and permanent seeding, mulching, earth dikes, silt fencing, straw-bale barriers, sediment basins, and mesh sheeting for ground cover.

Waste Management

Green remediation practices for waste management encourage consumers to consider lifecycle cost (including natural resource consumption) of products and materials used for remedial activities. BMPs build on requirements set by municipal or state agencies and those formalized in various construction and operating permits. A site management plan should include waste planning practices that apply to all cleanup and support activities. For sites involving construction and demolition or requiring diversion of landfill waste, stakeholder collaboration plays a significant role in sustainable cleanup.

BMPs for waste management during site cleanup are borrowed from the construction industry. Demolition concrete, for example, is often reused onsite as road base, fill, or other engineering material. Reducing and recycling debris such as concrete, wood, asphalt, gypsum, and metals helps to:

- Conserve landfill space,
- Reduce the environmental impact and cost of producing new materials, and
- Reduce overall project expenses through avoided purchase and disposal costs.

Waste management practices should consider every opportunity to recycle land-clearing debris, cardboard, metal, brick, concrete, plastic, clean wood, glass, gypsum wallboard, carpet, and insulation. Site preparation can include early confirmations with commercial haulers, deconstruction specialists, and recyclers. A convenient and suitably sized area should be designated onsite for recyclable collection and storage. Requirements for worker use of cardboard bailers, aluminum can crushers, recycling chutes, and sorting bins will facilitate the waste management program. In addition, stakeholders can help identify local options for material salvage that may include donation of materials to charitable organizations such as Habitat for Humanity. To document BMPs, site managers are encouraged to track the quantities of waste that are diverted from landfills during remediation.

Green waste management practices rely on recycling, reusing, and reclaiming materials to the greatest extent possible. [31]

Investigation derived waste such as drilling fluids, spent carbon, and contaminated personal protection equipment must be appropriately contained and stored outside of general recycling or disposal areas. Preference should be given to building- and equipment-cleaning supplies with low phosphate and non-toxic content.

SECTION 4: ENERGY AND EFFICIENCY CONSIDERATIONS

Energy requirements constitute a core element of green remediation. Significant reductions in fossil fuel consumption during treatment processes can be achieved through (1) greater efforts to optimize treatment systems, and (2) use of alternative energy derived from natural, renewable energy sources. "Active energy" systems use external energy to power mechanical equipment or otherwise treat contaminated media. These systems typically consume high quantities of electricity, and to a lesser extent natural gas, although duration of peak consumption varies among cleanup technologies and application sites. In 2007, approximately 70% of the U.S. electricity supply was generated by fossil fuel-fired plants.

CO_2 is one of several gases with potential to contribute to climate change. CO_2 is produced from a variety of sources including fossil fuel combustion and industrial process emissions. Electric power production is the largest source of CO_2 emissions in the U.S. energy sector, representing approximately one-third of the total.

EPA's Office of Solid Waste and Emergency Response (OSWER) is analyzing the extent of energy use, CO_2 emissions, and energy cost of technologies used to treat contaminated media at NPL sites. The analysis will help the Agency to:

- Establish benchmarks regarding the energy consumption of technologies with high energy demand,
- Examine operational and management practices typically used to implement these technologies, and
- Identify methods for reducing energy consumption during treatment processes and optimizing the systems.

The most frequently used energy-intensive treatment technologies used at NPL sites are pump-andtreat (P&T), thermal desorption, multi-phase extraction, air sparging, and soil vapor extraction (SVE). Using data from cost and performance reports compiled by the Federal Remediation Technologies Roundtable and other resources, OSWER estimates that a total of more than 14 billion kilowatt-hours (kWh) of electricity will be consumed through use of these five technologies at NPL sites from 2008 through 2030 (Table 2).

Table 2. Technologies used for Superfund cleanups often involve energy intensive components such as ground water extraction pumps, air blowers, or ultraviolet lamps (U.S. EPA/OS WER, 2008(b)

Technology	Estimated Energy Annual Average ($kWh*10^3$)	Total Estimated Energy Use in 2008-2030 ($kWh*10^3$)
Pump & Treat	489,607	11,260,969
Thermal Desorption	92,919	2,137,126
Multi-Phase Extraction	18,679	429,625
Air Sparging	10,156	233,599
Soil Vapor Extraction	6,734	154,890
Technology Total	***618,095***	***14,216,209***

DOE estimates that 1.37 pounds of CO_2 are emitted into the air for each kWh of electricity generated in the United States. Accordingly, use of these five technologies at NPL sites in 2008 through 2030 is anticipated to indirectly result in CO_2 emissions totaling nearly 9.2 million metric tons (Table 3) (U.S. EPA/OSWER, 2008(b)).

Based on the average electricity cost of $0.091 4/kWh in December 2007, consumption of fossil fuel energy at NPL sites during operation of these five technologies is anticipated to cost over $1.4 billion from 2008 through 2030. Use of these technologies under other cleanup programs such as RCRA, UST, or brownfields could produce similar results. Trends in the use of active energy treatment systems often vary among the various cleanup programs due to the type and extent of contamination and cleanup practices commonly encountered within each program.

General assumptions used in these estimates are dependent on and sensitive to factors such as site size or setting. The estimates do not include variable demands of additional electricity consumed during site investigations, field trials, remedy construction, treatment monitoring, and other activities. The Agency's online Power Profiler can help estimate air emissions attributable to electricity consumption at specific sites based on geographic power grids. [32]

Table 3. Estimated CO_2 emissions from use of five types of cleanup technologies at NPL sites over 23 years are equivalent to operating two coal-fired power plants for one year. [26]

Technology	Estimated CO_2 Emissions Annual Average (Metric Tons)	Total Estimated CO_2 Emissions in 2008-2030 (Metric Tons)
Pump & Treat	323,456	7,439,480
Thermal Desorption	57,756	1,328,389
Multi-Phase Extraction	12,000	276,004
Air Sparging	6,499	149,476
Soil Vapor Extraction	4,700	108,094
Technology Total	***404,411***	***9,301,443***

Optimizing Energy Intensive Systems

Significant reductions in natural resource and energy consumption can be made through frequent evaluation of treatment system efficiencies before and during operations. Opportunities to optimize systems and integrate high performance equipment begin during feasibility studies, when potential remedies are evaluated and the most appropriate and cost-effective cleanup technology is selected. In accordance with green remediation strategies, feasibility studies could include comparison of the environmental footprint expected from each cleanup alternative, including GHG emissions, carbon sequestration capability, and water drawdown (lowering of the water table or surface water levels).

The subsequent design phase involves planning of the selected technology's engineering aspects such as equipment sizing and integration. Energy consumption of remediation technologies ranges considerably, from soil excavation that requires virtually no mechanical integration or electrical power, to treatment trains involving media extraction and aboveground exposure to a series of electrically driven physical or chemical processes. In contrast to a "bottom up" approach, most cleanup technologies are designed through a series of equipment specifications requiring adjustment when components are integrated. Project solicitations for equipment and services should contain specifications regarding product efficiency, reliability, fuel consumption, air emissions, water consumption, and material content.

Selection of equipment and service providers must meet a project's performance and cost requirements, giving preference to products and user techniques working together to reduce environmental footprints.

Equipment and vendor selection can maximize use of alternative fuel and renewable energy sources. Where alternatives are currently unavailable or infeasible, designs can document the project's baseline energy demand for future reconsideration. Energy efficiency can be gained relatively simply by techniques such as insulating structural housing and equipment used to maintain certain process temperatures; installing energy recovery ventilators to maintain air quality without heat or cooling loss in treatment buildings; and weather-proofing system components that are exposed to outside elements. Electronic data systems for controlling and monitoring operations also provide significant opportunity to conserve energy, particularly in the multi-step processes commonly used for P&T. EERE has identified specific opportunities to identify and quantify energy efficiencies that might occur during pumping operations. Inefficiency symptoms include use of throttle-valve controls, cavitation noise or damage, continuous pumping to support a batch process, open bypass or recirculation lines, and functional changes of a system. [33]

Treatment system designs also should compare the environmental footprint left by alternate methods of managing process water, whether through re-injection to an aquifer, discharge to surface water, or pumping to a publicly owned wastewater treatment plant. Effective designs maximize every opportunity to recycle process fluid, byproducts, and water; reclaim material with resale value; and conserve water through techniques such as installation of automatic shut-off valves. To reduce impacts on water quality, construction designs can follow LID practices helping to infiltrate, evapotranspire, and re-use stormwater runoff in ways mirroring the site's natural hydrology.

Green remediation relies on maximizing efficiencies and reducing natural resource consumption throughout the duration of treatment. Upon process startup, tests are conducted to ensure the system is functioning as designed. For a technology such as in situ chemical oxidation, testing primarily involves ensuring that an injected material is reaching the target treatment zone. For a complex multi- contaminant P&T system, however, numerous tests are conducted to ensure that flow rates for each process step are appropriate and that equipment is properly sized.

Remedial system evaluations (RSEs) provide examples of BMPs already in place. EPA is conducting RSEs for operating P&T systems at Superfund-lead sites to:

- Indicate whether the original monitoring or treatment system design is fully capturing the target contaminant plume,
- Determine whether new monitoring or extraction wells are needed,
- Recommend specific modifications to increase system performance and efficiency, and

- Obtain cost savings from direct optimization or project management improvements. [34]

RSEs often find that energy intensive equipment such as pumps and blowers are oversized or set at operating rates or temperatures higher than needed, resulting in excess energy consumption (U.S. EPA/OSWER, 2002). Evaluations such as these also help to remove redundant or unnecessary steps in a treatment process, consider alternate discharge or disposal options for treated water or process waste, and eliminate excess process monitoring.

Standard operating procedures for treatment systems should include frequent reconsideration of opportunities to increase operational efficiencies. System optimization should carry forward to longterm operation and maintenance (O&M) programs that ensure system components are performing as designed. Poorly operating or broken equipment should be repaired immediately to avoid treatment disruption and energy waste.

Profile: Havertown PCP Site, Havertown, PA

Cleanup Objectives: Remediate shallow ground water containing metals, chlorinated volatile organic compounds (VOCs), benzene, and dioxins/furans

Green Remediation Strategy: Conducted RSE evaluation of a 12-acre treatment area encompassing

- Four recovery wells
- One collection trench
- A pre-treatment system to break oil/water emulsion, remove metals, and remove suspended solids in extracted ground water
- An aboveground system employing three 30-kW ultraviolet/oxidation (UV/OX) lamps, a peroxide destruction unit, and two granular activated carbon units to destroy or remove organic contaminants

Results:

- Removed two UV/OX lamps from the treatment line, based on RSE recommendations
- Reduced annual operating costs by $32,000, primarily due to lower electricity consumption
- Continues to meet cleanup criteria for ground water

Property End Use: Undetermined

Subsurface remediation generally changes dynamics of the natural system as well as distribution of contaminants. Changes might occur slowly, not becoming evident for several years. Periodic RSE helps to identify any subsurface changes, prompting modification to long-term treatment operations. Many years of P&T operations, for example, could change dynamics of plume behavior to the point where an outside extraction well that originally pumped contaminated water is later capturing clean water. In this case, shutdown of the extraction well will result in significant energy and cost savings.

Most remedies for soil and sediment (*in situ* oxidation, thermal treatment, and solidification/stabilization) are short-term in nature but require continual optimization throughout operations. Optimization of a biological system ensures that geochemical conditions such as reduction/oxidation, electron donor availability, and oxygen content are maximized.

In contrast to other soil and sediment technologies, SVE treatment results in contaminant loading that is initially high but decreases over time, prompting the need for frequent system modifications. Key opportunities for SVE optimization include (1) determining if any well in a manifold system is not contributing contaminants, and if so, taking the well offline, (2) operating pulsed pumping during off-peak hours of electrical demand, as long as cleanup progress is not compromised, and (3) considering alternative technologies with lower cost and energy intensity once the bulk of contamination is removed. The EPA, U.S. Air Force Center for Engineering and the Environment, Federal Remediation Technologies Roundtable, and Interstate Technology and Regulatory Council continue to develop tools such as checklists and case studies to help project managers optimize cleanup systems for all environmental media. [35-38]

Integrating Renewable Energy Sources

Incorporation of alternative, renewable energy sources into site cleanup may reduce a project's carbon footprint while offering other benefits:

- Hedge against fossil fuel prices, with the potential for near- and long-term cost savings,
- Lower demand on traditional energy sources,
- Reduced need for emission controls related to onsite fossil fuel consumption, and
- Opportunities for new energy markets and job creation when combined with site revitalization.

Renewable energy sources can be used to meet partial or full demand of a treatment system. When meeting partial demand, a renewable energy system can be designed to power one or more specific mechanical components or to generally supplement grid electricity supplied to the entire treatment process. EPA's *Green Power Equivalency Calculator* could be used to better understand and communicate the environmental benefits of directly or indirectly using electricity produced from solar, wind, geothermal, biogas, biomass, and low-impact small hydroelectric sources, otherwise known as "green power. " [39]

Energy alternatives already available for remediation and revitalization include solar, wind, landfill gas, and waste-to-energy sources. Emerging technologies such as geothermal and tidal power also could be used for site-wide applications or as means to optimize treatment system components. Potential integration of renewable energy sources considers:

- Natural resource availability, reliability, and seasonal variability,

- Total energy demand of the treatment system,
- Proximity to utility grids, and associated cost and time needed to connect to the grid,
- Back-up energy sources for treatment or safety,
- Cost tradeoffs associated with cleanup duration and economy of scale, and
- Long-term viability and potential reuse.

Renewable energy industries estimate a current renewable energy capacity of 550-770 gigawatts in the United States, with growth sufficient to meet at least 25% of the country's electricity needs by 2025. (ACORE/ABA, 2008)

Renewable energy provides significant opportunities at sites that require long-term treatment, are located in remote areas, or involve energy intensive technologies such as P&T. Renewable energy systems can operate independently without connection to a utility grid (off-grid) or as interconnected systems tied to the utility power grid (inter-tie). Energy management tools can be used to monitor supply and demand, automatically shutting off or initiating grid power as desired. Hybrid systems combining capability of two or more renewable resources often provide the most efficient and cost- effective option in rural areas or to achieve total energy independence.

Off-grid systems are best suited to mechanical or infrastructure components with low or intermittent energy demands such as small pumps, communication systems, or the interior of small buildings. Cost effectiveness of off-grid systems significantly increases at remote sites, where extension of utility lines might be cost prohibitive or otherwise infeasible due to difficult access. As in all optimized engineering systems, effective renewable energy systems include climate control measures to minimize energy loss throughout the mechanical network.

Interconnection of renewable energy systems with the utility grid allows use of utility power when availability of a natural resource is low, without disruption to site cleanup operations. Excess energy produced by a small renewable energy system could be stored in batteries until needed or transferred to the grid for consumption by other users. Most states now require electric utilities to offer net metering, a service that enables renewable energy generators to receive utility consumption credit. The amount of excess electricity transferred to the grid could be directly measured through installation of an additional meter or generally monitored through visual observation of the primary utility meter "spinning backward." DOE's National Renewable Energy Laboratory (NREL) is working with other government agencies and private industry to develop consistent standards for grid interconnection, system engineering, and power production market rules.

Capital costs for renewable energy systems continue to decrease as technologies advance and as demand steadily increases but might prohibit their use in some projects. Costs can be lowered by taking advantage of federal and state rebates or tax credits or shared through reuse of equipment in other cleanup projects. Project decision-makers are encouraged to capitalize on 10-year renewable energy incentives now available to help capture long-term savings while strengthening community economics.

Profile: Former St. Croix Alumina Plant, St. Croix, VI

Cleanup Objectives: Recover hydrocarbons from ground water at a RCRA site

Green Remediation Strategy: Uses a hybrid system employing solar and wind energy

- Began operating four wind-driven turbine compressors in 2002 to drive compressed air into hydraulic skimming pumps
- Installed three 55-watt photovoltaic panels in 2003 to power some recovery wells
- Added three 110-watt photovoltaic panels and two wind-driven electric generators in 2006 to power a total of nine submersible total-fluid pumps and the fluid-gathering system
- Recycles recovered petroleum product by transfer to an adjacent oil refinery for use as feed stock

Results:

- Recovered 228,000 gallons of free- product oil (approximately 20% of the estimated volume) by the end of 2006 - Avoids offsite transfer and disposal of petroleum product

Property End Use: Industrial operations (U.S. EPA/OSWER, 2008(c))

Increasing numbers of regional partnerships are forming to help property owners install large utility-grade systems that can meet energy demands of onsite operations such as remediation, while receiving production tax credit and allowing sale of excess energy to the utility at wholesale price. Another mechanism is the power purchase agreement, which enables owners of large properties to lease land to a utility for installation and operation of a renewable energy network (typically solar or wind systems) while purchasing electricity at a considerably lower rate. These partnerships add to renewable energy portfolios maintained by state agencies and authorized utilities to help meet national goals. Accordingly, new generators of renewable energy are actively solicited by states working to meet the goals of renewable portfolios. [40]

Solar Energy

Solar energy can be used in site cleanups through one or more methods involving photovoltaics (PV), direct or indirect heating and lighting systems, or concentrating solar power. PV technology easily lends itself to applications involving remote locations, a need for portability, or support for long-term treatment systems. This technology is already in place or under design at numerous sites.

PV cells consist of absorbing, semiconducting material that converts sunlight directly into electricity. Typically, about 40 PV cells are combined to form a module, or panel. Approximately ten of the modules are combined on a flat-plate PV array that might range several yards in size. An array could be mounted at a fixed angle facing south, or on a tracking device following the sun to allow maximum capture of sunlight over the course of a day. Six to 12 modules might meet all or part of a treatment system with low energy demand. In contrast, 1 0-20 arrays could be needed to power systems on the order of a small industrial facility or hundreds of arrays can be interconnected to form a single system.

Profile: BP Paulsboro, Paulsboro, NJ

Cleanup Objectives: Remove petroleum products and chlorinated compounds from surface and ground water near a Delaware River port

Green Remediation Strategy: Uses a solar field to power P&T system extracting 300 gallons of ground water per minute

- Installed a 275-kW solar field encompassing 5,880 PV panels in 2003
- Uses solar energy to operate six recovery wells including pump motors, aerators, and blowers
- Transfers extracted ground water into a biologically activated carbon treatment system

Results:

- Supplies 350,000 kWh of electricity each year, meeting 20-25% of the P&T system energy demand
- Eliminates 571,000 pounds of CO_2 emissions annually, equivalent to avoiding consumption of 29,399 gallons of gasoline
- Prevents emission of 1,600 pounds of sulfur dioxide and 1,100 pounds of nitrogen dioxide each year
- Provides opportunity for reuse and expansion of the PV system, with potential capital cost recovery if integrated into site reuse

Property End Use: Port operations (U.S. EPA/OSWER, 2007(a))

Use of solar energy at the Pemaco Superfund site in Maywood, CA, demonstrates the flexibility and capability of solar technology in helping to meet energy demands of above-ground treatment operations. Four PV panels with a total generating capacity of 3 kW were installed on the existing building, which houses a soil and ground water treatment system employing high-vacuum pumps, controls for electrical resistance heating, a granular activated carbon unit, and a high- temperature flameless thermal oxidizer. The PV system contributes a total of 375 kWh of electricity to the building operations each month, avoiding more than 4,300 pounds of CO_2 emissions per year. After the first nine months of operation, solar energy had generated enough power to cover one month of the building's electricity expenses for system controls and routine operations. Payback for PV capital costs is estimated at one year.

Aboveground treatment processes also can use solar thermal methods. These methods employ solar collectors such as engineered panels or tromb walls to absorb the sun's energy, providing low- temperature heat used directly for space heating. In contrast, solar water heaters use the sun to directly heat water or a heat-transfer fluid in collectors. Industrial-grade solar heaters can be used to provide hot water and hot-water heat for large treatment facilities.

Passive (non-mechanical) methods also could be used to heat treatment buildings, potentially reducing structural energy consumption by up to 50%. Buildings can be designed to include large spans of windows with southern exposure or constructed of materials with high mass value (high absorbency but slow heat release). Passive solar designs also include natural ventilation for cooling. Daylighting of treatment buildings can be enhanced through installation of conventional skylights or smaller "tubular skylights" constructed of reflective material. Also, parabolic solar collectors could be used to supplement electricity demands of fiber optic systems for treatment monitoring or data transfer.

The potential for using active or passive solar energy to meet the energy demands of treatment processes throughout the year can be calculated using the site's estimated insolation. An insolation value indicates the rate at which solar radiation is delivered to a unit of horizontal surface. Insolation values indicate radiation reflection or absorption by (1) flat-plate collectors facing south at fixed tilt (Figure 4), (2) single-axis (north/south) flat-plate collectors tracking from east to west, (3) two-axis flat-plate collectors tracking the sun in both azimuth and elevation, or (4) concentrating collectors using multiple axes to track direct solar beams. Technical assistance and more information is available from NREL and the American Solar Energy Society to help site managers determine whether the energy demands of site remediation as well as anticipated reuse could be met by solar resources. [41, 42]

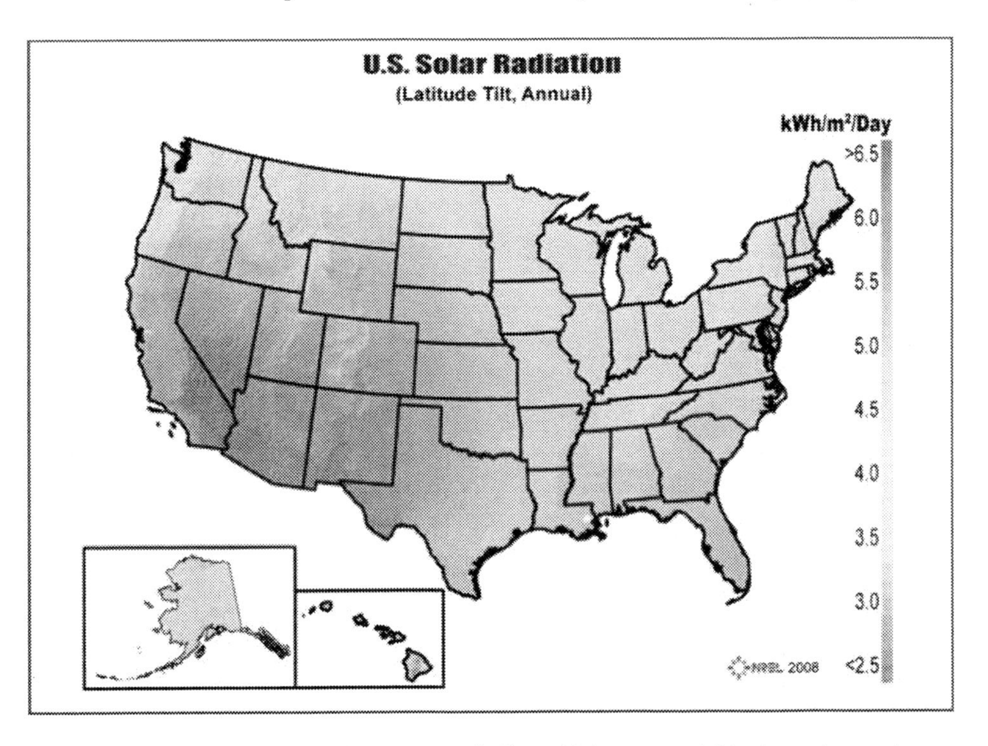

Figure 4. Estimates of U.S. annual solar resources indicate highest potential in the outhwest; in areas with lowest potential, resources remain equivalent to those of Germany, where solar energy is used routinely across business sectors. (U.S. DOE NREL, 2008(a))

Concentrating solar power (CSP) systems provide significant opportunities at large sites undergoing cleanup and revitalization. CSP systems use reflective materials such as mirrors or parabolic troughs to concentrate thermal energy driving an electricity generator, or concentrated PV technology to directly provide electrical current. Large-scale CSP systems are under consideration at sites in southwestern portions of the United States.

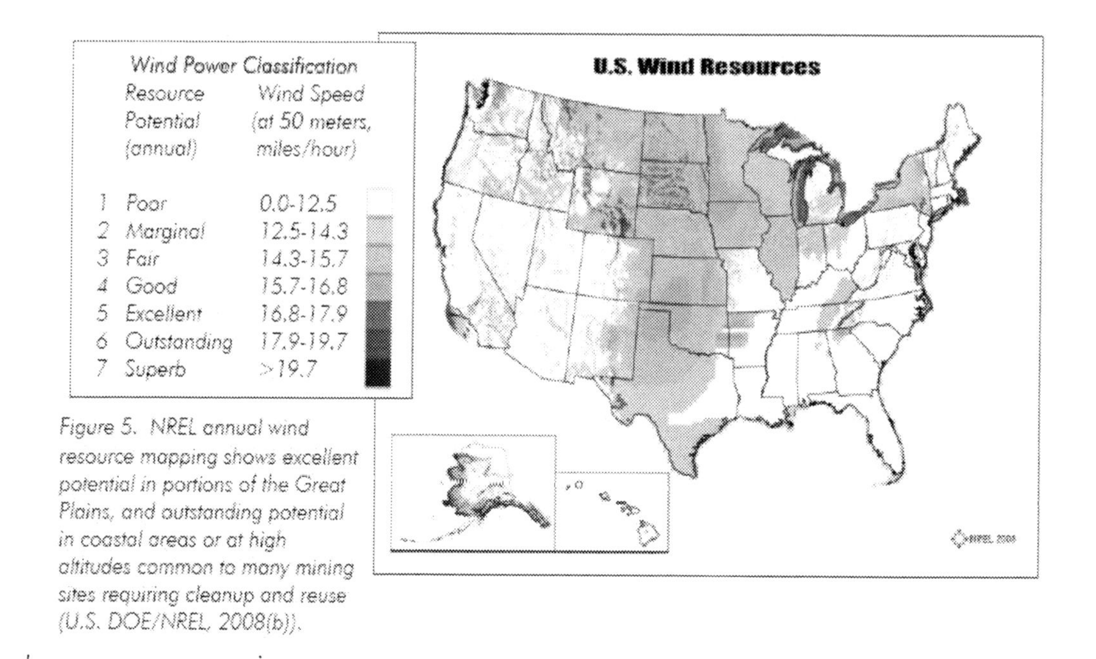

Figure 5. NREL annual wind resource mapping shows excellent potential in portions of the Great Plains, and outstanding potential in coastal areas or at high altitudes common to many mining sites requiring cleanup and reuse (U.S. DOE/NREL, 2008(b)).

Figure 5. NREL annual wind resource mapping shows excellent potential in portions of the Great Plains, and outstanding potential in coastal areas or at high altitudes common to many mining sites requiring cleanup and reuse (U.S. DOE/NREL, 2008(b))

Wind Energy

Determining the potential for using wind energy to meet energy demands of a cleanup requires a wind resource assessment. The assessment involves collection of climatic data from an onsite or local weather station over the course of one year, although DOE's wind resource data might be sufficient for small applications on relatively flat terrain. Wind speed is critical but wind shear and turbulence intensity also impact assessment results. Generally, the amount of power available by wind is proportional to the cube of wind speed; for example, a two-fold increase in wind speed increases the available power by a factor of eight. Wind energy is best suited to resource areas categorized as "Class 3" or higher on DOE's scale of 1 -7 (Figure 5). [43]

Results of the wind resource assessment are compared to the cleanup's anticipated energy demand to determine whether wind energy would meet full or partial demand. Demands of low energy components such as small generators might be met by wind speeds of 6 miles per hour (mph), while activities such as ground water pumping generally require a wind speed above 9 mph. At sites with wind speeds averaging 12 mph, a small 10-kW wind turbine can generate approximately 10,000 kWh annually (equivalent to avoiding CO_2 emissions resulting from consumption of 882 gallons of gasoline).

In addition to wind speed, output of a wind turbine significantly depends on a turbine's size. Most small turbines consist of a rotor (encompassing the gearbox and blades) with diameters of less than 1 0 feet, mounted on towers 80-120 feet in height. Due to the low number of moving parts, most small turbines require little maintenance and carry an estimated lifespan of 20 years. Small systems cost $3,000-5,000 for every kilowatt of generating capacity, or approximately $40,000 for a 10-kW installed system.

Treatment systems requiring compressed air could be powered by wind-driven electric generators. This type of generator employs a small turbine or windmill to capture, compress, and direct air to equipment such as hydraulic pumps. The generator typically is designed to allow blade rollup and repositioning during excessive wind, and can easily be lowered to the ground for routine maintenance.

Increasing numbers of communities are examining opportunities for integrating renewable energy production into a contaminated site's long-term viability and reuse. Site revitalization involving production of electricity for utility distribution requires installation of co-located utility-scale (100- kW or more) turbines to form a wind farm (wind power plant). A wind farm is best suited to areas with wind speeds averaging at least 13 mph. A one-megawatt (MW) turbine can generate 2.4-3 million kWh annually; a 5-MW turbine can produce more than 15 million kWh annually. Capital and installation costs range according to factors including economy of scale and site-specific conditions such as terrain.

Profile: Former Nebraska Ordnance Plant, Mead, NE

Cleanup Objectives: Remove trichloroethene and destroy explosives in ground water

Green Remediation Strategy: Uses a 10-kW wind turbine to power ground water circulation wells for air stripping and UV treatment

- Calculated a total demand of 767 kWh each month for the circulation wells
- Determined electricity demand could be met by site conditions including wind speed of 6.5 meters/second

Results:

- Provides sufficient energy for continued trichloroethene removal and explosives destruction by the aboveground treatment system during grid inter-tie operation
- Reduces consumption of utility electricity by 26% during grid inter-tie operation
- Decreases CO_2 emissions by 24-32% during off-grid operation of the system's 230-volt submersible pump
- Returns surplus electricity to the grid for other consumer use
- Results in no observable impacts on wildlife
- Provides electricity cost savings expected to total more than $40,000 over the next 15 years of treatment
- Estimated that cost recovery time for turbine capital and installation could be cut in half by improved freeze-proofing of wells

Property End Use: Continued agricultural research and development, residential, and commercial use (U.S. EPA/OSWER, 2007(b); University of Missouri- Rolla, 2005)

Integration of utility-scale energy production in site reuse considers efficiencies as well as economic factors. Commercial wind turbines average a mechanical and electric conversion efficiency of approximately 90%, and an aerodynamic efficiency of approximately 45%. In contrast, the average efficiency of electricity generating plants in the United States averages approximately 35%; over two-thirds of the input energy is wasted as heat into the environment.

Over the last 20 years, the cost of electricity from utility-scale wind systems has dropped more than 80%, from an earlier high of approximately 80 cents per kWh. With the use of production tax credits, modern wind power plants can generate electricity for 4-6 cents/kWh, which is competitive with the cost of new coal- or gas-fired power plants.

Design of a small wind system includes consideration of horsepower across the entire system to maximize efficiency. Ground water pumps, for example, typically operate at 50% efficiency, while turbine efficiency typically exceeds 90% and grid efficiency averages about 91% (U.S. DOE/EIA, 2007). Efficiency can be enhanced by grid interconnection allowing higher start-up current to be drawn from the grid and by avoiding the need for storage batteries.

Wind plants typically are designed in modules allowing for addition or subtraction of individual turbines as electricity demand changes. Construction of a 50-MW wind farm can be completed in six months, beyond the initial 12-18 months commonly needed for wind measurements and construction permits.

For maximum efficiency, installation locations should be sufficiently distant from trees or buildings that potentially reduce speed of wind entering the turbine. Selection of turbine sites also considers potential impacts on sensitive environments made by turbine noise (commonly compared to a domestic washing machine) and public perceptions regarding aesthetics of turbine sizes. A typical 1 00-kW turbine contains a rotor approximately 56 feet wide, while rotor width of a 1 ,650-kW turbine averages 233 feet. Height of a utility-scale tower ranges according to site conditions but generally is similar to rotor width.

EERE estimates wind energy is the fastest growing energy generation technology, expanding 30-40% annually. NREL and the American Wind Energy Association offer technical assistance on evaluating and implementing wind systems. [44]

Landfill Gas Energy

Landfill gas (LFG) generated through decomposition of solid waste provides a potential source of energy at numerous sites across the country with abandoned or inactive landfills. LFG typically contains about 50% CO_2 and 50% CH4. LFG-to-energy systems use extraction wells to capture gas before it enters the atmosphere or is burned as part of the landfill management process. Captured gas can be converted to an alternate fuel, to electricity for direct use, or to both electricity and thermal energy (co-generated heat and power, or CHP) for dedicated mechanical operations. [45]

Conversion of LFG to electricity is possible through a number of technologies, depending on the scale of generation. Proven technologies include microturbines, internal combustion engines, gas turbines, external combustion engines, organic Rankine cycle engines, and fuel cells. Microturbines range in power from 30 kW to 250 kW (not exceeding 1 MW), internal combustion engines range from 1 00 kW to 3 MW; and gas turbines range from 800 kW to 1 0.5 MW. Although combustion of LFG converts CH_4 to CO_2, the global warming potential of methane is 23 times higher than that of carbon dioxide. Increasing numbers of LFG applications involve development of aerobic digesters that rely exclusively on anaerobic bacteria to break down organic substances.

Effective design of an LFG energy system includes adequate conditioning that ensures converted gas is free of vapor and remaining contaminants or impurities, and operational practices that minimize liquid waste streams. Performance and lifespan of a system depend on long-term availability and reliability of the methane as an energy resource.

LFG energy systems benefit from economy of scale. For example, EPA estimates that the total installed cost for an LFG microturbine project falls from $4,000- 5,000 per kW for a small (30-kW) system to a cost of $2,000-2,500 per kW for systems rated 200 kW and higher. As of early 2007, 424 LFG energy projects operated in the United States, producing a total 1,195 MW of electricity. EPA estimates that an additional 560 landfills hold potential for converting LFG to productive use, with a total production potential of 1 ,370 MW of electricity. [46] This technology brings significant potential for reducing GHG emissions from landfills. The community of Shippensburg, PA, for example, anticipates that operation of its 6.4-MW LFG electricity- generating system will prevent emission of 39,000 tons of CO_2 each year (an equivalency of one coal-fired power plant generating electricity for nearly 660,000 homes).

Profile: Operating Industries, Inc. Landfill, Monterey Park, CA

Cleanup Objectives: Remediate soil and ground water contaminated by a 145-acre inactive landfill

Green Remediation Strategy: Convert LFG to electric power for onsite use

- Installed six 70-kW microturbines in 2002 as part of the LFG collection system
- Converts a LFG flow rate of 5,500 standard cubic feet per minute, with a CH_4 content of approximately 30%
- Returns microturbine emissions to the existing gas treatment system to ensure contaminant removal

Results:

- Generates sufficient energy to meet approximately 70% of onsite needs including thermal oxidation, a 40-horsepower gas blower, refrigeration units, and air-exchange systems
- Saves up to $400,000 each year in grid- supplied electricity expenses

Property End Use: Commercial/industrial operations or open space, pending Superfund close-out (U.S. EPA/OSWER, 2007(a))

Waste-Derived Energy

Waste-to-energy (WTE) systems convert solid waste into electricity, or in some cases liquid waste to alternative fuel. Large sites undergoing remediation provide opportunities for local communities to consider reuse options involving WTE facilities as a means to:

- Reduce municipal landfill burdens posed by disposal of non-hazardous waste,
- Provide an alternative to onsite landfill construction,

- Procure a long-term source of renewable energy,
- Decrease export of waste from communities with little or no landfill capacity to other facilities, often in other states, and
- Provide employment opportunities.

An average municipal WTE facility emits 837 pounds of CO_2 per megawatt hour; in contrast, coal, oil, and natural gas facilities emit over 2,000, 1,600, and 1,1 00 pounds of CO_2 per megawatt hour, respectively (Solid Waste Association of North America, 2005; U.S. EPA, 2008 online). DOE's Energy Information Administration estimates that a total of 299 trillion British thermal units of energy were consumed by combustion of municipal solid waste in 2005 (U.S. DOE/EIA, 2008). Conversion of heat produced during this process is used increasingly to produce electricity. For example, Lee County, FL, recently expanded its existing 4 million-ton WTE combustion system to process an additional 636 tons of municipal waste each day, resulting in production of an additional 1 8 MW of electricity.

Capital and operating costs for WTE facilities are significantly higher than conventional landfill costs and typically are covered through local bonds. To ensure long-term viability, WTE facilities rely on an infrastructure that guarantees a minimum quantity of incoming solid waste. The estimated lifespan of a WTE facility is 40 years. [47]

Developing and Evolving Energy Sources

Green remediation relies on novel applications of emerging technologies within the context of site cleanup. Technologies for producing energy from previously untapped renewable resources are quickly moving from research facilities into the field, significantly increasing the options available for site revitalization. Integrated planning for site cleanup and reuse borrows principles used in this "next generation" of renewable energy technologies but also resurrects past methods for obtaining energy from natural resources such as "old-fashioned" windmills or small-scale hydropower.

Geothermal power is energy generated by heat stored beneath the earth's surface, whether stored in shallow ground or in water and rock at depths extending several miles below ground surface. Temperatures in ground water and rock at subsurface depths up to 1 0 feet remain relatively constant at 50-60°F, bringing potential for geoexchange systems to be used in remediation. Aboveground treatment methods can use this energy directly through installation of air exchange pumps to heat or cool building interiors. Heat removed from indoor air also could be used to elevate the temperature of water required in a treatment process.

In contrast to heat exchange, new technologies for cold energy storage could help cool treatment processes and structures at sites located adjacent to cold water reservoirs. For example, the Halifax Regional Municipality began construction of a $3 million energy system retrofit in 2007 to meet peak air conditioning needs of buildings along the waterfront in Dartmouth, Nova Scotia, Canada. The system employs a borehole exchanger drawing cold air from 1 00 holes extending 600 feet below ground surface to tap energy from subsurface rock mass.

Geothermal resources at greater subsurface depths could be considered to generate electricity for long-term cleanup as well as potential sale. Geothermal power plants currently coming into operation in western states tap reservoirs of water with temperatures of 1 07-1 82°F, which are considerably lower temperatures than needed in past production. New plants operate at lower cost and greater efficiency, and emit significantly less CO_2 than fossil fuel plants (less than 1 00 pounds per megawatt hour). Potentially adverse environmental problems posed by geothermal energy production include process operations requiring deep subsurface drilling and condensed steam re-injections to draw additional heat; changes in geological stability of a region; and decreasing temperatures of water reservoirs over time. [48]

Tidal energy could provide opportunities at coastal sites undergoing long-term treatment. Although ocean tide has not yet been tapped for remediation purposes, small-scale variations relying on the flow of ground water and surface water are under evaluation. For example, DOE's Savannah River National Laboratory field tested a passive siphoning system using a synthetic tube to induce ground water flow from a contaminated aquifer into a treatment cell containing reactive material. After passing through the treatment cell, water discharges to nearby surface water. System recharge, when needed, can be accomplished easily through use of a solar powered vacuum pump to remove gas bubbles. This technology provides a passive, *in situ* alternative to P&T systems and could be used to improve performance of other low energy technologies such as permeable reactive barriers.

Adaptations of conventional treatment technologies can take advantage of energy produced by other earth processes. Passive bioventing or passive SVE rely on natural venting cycles of the subsurface to create atmospheric pressure differences capable of inducing air flow (barometric pumping) for subsurface removal of nonchlorinated hydrocarbons. Effectiveness is enhanced through simple air- control equipment such as one-way valves preventing flow of air into venting wells. The U.S. Air Force Center for Engineering and the Environment is evaluating long-term efficiency of pressure-driven systems at numerous sites, including Hanford, WA, and Hill Air Force Base, UT. Pressure-driven systems do not require mechanical pumps or electrical blowers to draw volatile contaminants from soil and provide a low-cost approach for remediation polishing following use of energy-intensive remediation technologies. Applications are limited to sites with substantial swings in barometric pressures and are most effective under aerobic conditions in shallow, unsaturated soil. Passive pressure systems commonly require more venting wells than conventional systems and often require longer time to achieve cleanup goals. [49]

The Savannah River National Laboratory is testing low power (20-40 watt) SVE systems powered by small PV modules, wind generators, or batteries. Pumps used in these applications are small and relatively unobtrusive (typically four by three inches in size) but might need replacement after one year of operation. Use of low power SVE is limited to long-term remediation polishing. [50]

Low Energy Systems

Passive energy remediation systems use little or no external energy to power mechanical equipment or otherwise treat contaminated environmental media. These systems commonly involve technologies such as bioremediation, phytoremediation, soil amendments, evapotranspiration covers, engineered wetlands, and biological permeable reactive barriers. Cleanup strategies can combine elements of these technologies to achieve novel hybrid systems, paving the way for yet more innovative applications.

To maximize remediation sustainability, passive energy systems should operate in conjunction with other core elements of green remediation such as water conservation and waste minimization; rely on energy efficient equipment during construction and monitoring; and consider use of renewable energy sources for auxiliary equipment. As in all cleanup actions, selection and implementation of remedies relying on passive energy technologies must account for short- and longterm environmental and cost trade-offs. Passive systems often require more time than aggressive, active energy systems to meet cleanup goals.

Carbon sequestration is the removal from the emission stream of CO_2 or other GHG that would otherwise be emitted to the atmosphere. GHGs can be sequestered at the point of emission or removed from air, often referred to as carbon capture and storage. Emissions can be offset by enhancing carbon uptake in terrestrial ecosystems and subsequent carbon storage in soil. Vegetation serving as "carbon storage sinks" adds to the earth's net carbon storage. [51]

These systems can serve as the primary means for treating contaminated media or as secondary polishing steps once the effectiveness of more energy intensive systems begins to be outweighed by negative cost and environmental affects. Passive energy systems can increase terrestrial sequestration of CO_2 and other GHG, resulting in a "co-benefit" of site remediation. Monitoring and controls are required, however, to minimize potential for these systems to act as atmospheric CO_2 sources. For potential application in carbon offset programs becoming available in government and industrial sectors, systems must demonstrate permanence of atmospheric carbon sequestration as well as the amount of carbon being newly sequestered.

Passive energy systems inherently complement efforts to protect and restore ecological systems on contaminated lands, one of the core elements of green remediation. In addition to enhancing wildlife and vegetative habitat, ecological land use can provide features such as commercial riparian zones or recreational opportunities. Improved soil stability gained by ecological restoration also reduces erosion, slows and filters stormwater runoff, and reduces topsoil lost as dust during both remediation and reuse activities.

Enhanced Bioremediation

Enhanced bioremediation helps microorganisms degrade contaminants in soil, ground water, or sludge. In situ applications involve subsurface injection of microbial enhancing substrates, which results in minimal disturbance to land or ecosystems and little fuel consumption. *Ex situ* bioremediation involves disturbance to upper soil layers and requires more field activity but avoids offsite disposal of contaminated soil and associated consumption of vehicular fuel for transport. Depending on the selected technique, ex situ bioremediation can produce significant amounts of nutrient-rich material available for onsite or potentially commercial offsite applications.

Profile: Umatilla Army Depot, Hermiston, OR

***Cleanup Objectives*:** Treat 15,000 tons of soil contaminated with explosives such as trinitrotoluene (TNT) and cyclotrimethylenetrinitramine (RDX)

Green Remediation Strategy: Composted with locally obtained feedstock

- Used windrow techniques involving placement of soil in lengthy piles
- Periodically mixed soil with a mixture of cattle/chicken manure, sawdust, alfalfa, and potato waste
- Mixed soil with feedstock inside mobile buildings to control fumes and optimize biological activity

Results:

- Treated each 2,700-cubic-yard batch of soil in 10-12 days
- Destroyed contaminant byproducts or permanently bound the byproducts to soil or humus, achieving non-detectable concentrations of explosives
- Provided $150,000 potential revenue from sale of humus-rich soil
- Saved an estimated $2.6 million compared to incineration, a common alternative for explosives treatment
- Avoided significant fossil fuel consumption by an incinerator
- Avoided fuel costs and consumption associated with transporting soil to an offsite incinerator or transferring ash generated by an onsite mobile incinerator

Property End Use: Conversion under base realignment and closure
(U.S. EPA/OSWER, 1997)

In situ aerobic bioremediation typically is enhanced by injection of oxygen and/or moisture as well as compounds influencing media temperature and pH. The end product comprises primarily CO_2 and water. *In situ* anaerobic bioremediation processes typically are enhanced by injection of an electron donor substrate such as vegetable oil to promote suitable conditions for microbial growth. If the appropriate contaminant-degrading microbes are not present in sufficient quantity, additional microbes will be injected (bioaugmentation). Some applications targeting ground water create flow-through bioreactors or permeable reactive barriers constructed of organic material.

Ex situ bioremediation of soil may be conducted through a slurry process, whereby contaminated soil is excavated and mixed with water to suspend solids and provide contact with microorganisms. In contrast, solid-phase bioremediation involves placement of contaminated soil in a treatment cell or aboveground structure where it is tilled with water and nutrients. Land farming, biopiles, and composting are among the solid-phase bioremediation techniques producing enriched soil for potential use in landscaping and agriculture at revitalized sites. [52]

Ex situ enhanced bioremediation can play a significant role in green remediation by helping to rebuild organic content of soil, increase soil aeration, improve water infiltration, increase moisture retention, and stimulate vegetation growth. BMPs of green remediation include methods to control soil erosion and sediment transport through strategies such as topsoil stockpiling, installation of straw barriers, and placement of permeable ground cover to prevent soil compaction caused by heavy machinery. The practices also encourage air protection strategies such as use of clean fuel in on-road vehicles, retrofitting of diesel equipment, and minimal idling of heavy machinery.

Phytoremediation

Phytoremediation uses plants to remove, transfer, stabilize, or destroy contaminants in soil, sediment, and ground water. This technology encompasses all biological, chemical, and physical processes influenced by plants, including the root biomass (rhizosphere). Treatment mechanisms include:

- Phytoextraction (phytoaccumulation and phytotranspiration) involving contaminant uptake by plant roots and subsequent storage or transpiration of contaminants in plant shoots and leaves,
- Enhanced rhizosphere biodegradation, whereby contaminants break down in soil or ground water surrounding plant roots,
- Phytodegradation, whereby plant tissue metabolizes contaminants, and
- Phytostabilization, whereby plants produce chemical compounds to immobilize contaminants at the root/soil interface.

Profile: Carswell Golf Course, Fort Worth, TX

Cleanup Objectives: Biodegrade subsurface VOCs through reductive dechlorination and control contaminant migration

Green Remediation Strategy: Planted 660 cottonwood trees across 4,000 square meters in 1 996 to:

- Establish root biomass promoting activity of indigenous microbes
- Enhance transpiration of ground water through the trees, helping to control hydraulic gradient and down gradient migration of VOCs

Results:

- Produces virtually no process residuals
- Reduced VOC concentrations in ground water approximately 65% within four years after the plantings,
- Demonstrates increased treatment efficacy over time according to plant growth
- Incurred costs of only $2,100 for plants and $10,000 for irrigation
- Supported transfer of property to community as part of base closure, without disruption to ongoing activities

Property End Use: Recreation [U.S. EPA/OSRTI, 2005]

Plant communities used in phytostabilization can serve as significant carbon storage sinks. Carbon uptake during photosynthesis increases plant growth rate, in turn increasing biomass capability to capture and store atmospheric carbon. BMPs for phytoremediation rely on the use of native, noninvasive, and non-noxious plants. While selection of suitable plants is site-specific, vegetation with capability to treat contaminated soil or ground water includes common plants such as hybrid poplars, Bermuda grass, and alpine pennycress.

Phytoremediation systems can be constructed and maintained at low cost, depending upon site characteristics and goals, and require minimal equipment once installed.

LEED-based water efficiency goals for phytoremediation could include 50% use of non-potable water for irrigation, where needed. Methods to minimize water consumption include use of drip irrigation techniques, greywater reclaimed from industrial or small-scale potable water systems, and high efficiency equipment or climate-based controllers.

Phytoremediation can be used to treat organic compounds through the process of mineralization, and heavy metals or other inorganic compounds through the processes of accumulation and stabilization. The technology can be applied in situ to soil, sediment, or ground water. Applications involving no accumulation of contaminants (and associated disposal of plants) particularly complement land use that is dependent on bioversity, such as greenspace. [53]

Soil Amendments

Soil amendments are organic materials that can be applied in situ to enhance contaminant biodegradation by subsurface microorganisms and to decrease availability of metal contaminants. Soil amendments help restore degraded lands and ecosystems by:

- Improving water retention (resulting in enhanced plant growth and drought resistance) and other soil properties such as pH balance,
- Supplying nutrients essential for plant growth, including nitrogen and phosphorous as well as essential micronutrients such as nickel, zinc, and copper, and
- Serving as an alternative to chemical fertilizers that incur additional project costs and potentially introduce human health or environmental concerns.

In contrast to the quick release of nutritional elements following application of inorganic fertilizers, organic nutrients in soil amendments are released slowly, resulting in more efficient plant uptake and subsequent growth. Nutrients bound in organic matter also are less water soluble, rendering them less likely to leach into ground water or migrate as runoff into surface water. The process of applying soil amendments can be completed at a relatively low cost and often produces soil for use in site redevelopment. Applications must include precautions, however, to avoid potential nutrient- or metals-loading that contributes to nonpoint pollution of other environmental media.

"Biosolid recycling" of stabilized sewage sludge, which is increasingly used by municipalities as an alternative to incineration, provides a significant source of organic material needed to amend soil at hazardous waste sites. This approach converts organic wastewater treatment material into products for beneficial use such as bulk application in agriculture or pellets in commercial fertilizers. Generation and use of biosolids are subject to federal, state, and local requirements to ensure that treatment systems sufficiently sterilize organic material; excess field application is avoided; sufficient post-application time is allowed before plant harvesting; and metal content is within safe levels. [54]

Evapotranspiration Covers

Evapotranspiration (ET) covers are waste containment systems providing an alternative to conventional compacted-clay covers (caps) that might insufficiently prevent percolation of water downward through the cover to the waste. ET covers use one or more vegetated soil layers to retain water until it is transpired through vegetation or evaporated from the surface of soil. An ET cover also is known as a water balance cover, alternative earthen final cover, vegetative landfill cover, soil-plant cover, or store-and-release cover. These systems increase vegetative growth, help establish small wildlife habitat, and provide significant opportunities for CO_2 capture and sequestration.

Effective cover designs incorporate methods to control percolation and moisture buildup and to promote surface water runoff. ET covers rely on a soil layer's capacity for water storage, instead of engineered material with low hydraulic conductivity, to minimize percolation. Cover designs emphasize use of:

- Native vegetation to increase evapotranspiration, and
- Local soil to streamline construction, minimize project costs, and avoid fuel consumption associated with imported soil.

ET cover systems generally are constructed as monolithic barriers or capillary barriers. A monolithic cover (or monofill cover) uses a single vegetated layer of soil to retain water until it is either transpired through vegetation or evaporated from the soil surface. A capillary barrier cover system uses a similar clay layer typically underlain by sand or gravel to cause infiltrating water to wick at the layer interface.

Profile: Upper Arkansas River, Leadville, CO

Cleanup Objectives: Restore soil and ecosystems severely degraded by past mining activities conducted upstream

Green Remediation Strategy: Introduced biosolids and assorted soil amendments

- Applied 100 dry tons (pellets) of biosolids to each of 20 target acres along an 11-mile stretch of the river
- Mixed biosolids with lime to reduce soil acidity, consequently increasing plant viability and metal insolubility
- Seeded native plants and quick-growing rye grass
- Added compost and woody material as additional plant nutrients
- Added wood chips to reduce nitrogen (nutrient) leaching
- Covered amended soil with native hay to promote plant growth and seeding

Results:

- Revegetated denuded acreages
- Reduced concentrations and bioavailability of zinc and other metals through bioremediation, phytoremediation, and solubility reduction
- Neutralized soil to levels supporting healthier ecosystems
- Reduced soil erosion, river channel degradation, and property loss
- Reestablished communities of native plants such as white yarrow and tufted hairgrass

Property End Use: Agriculture and recreation

Costs for construction could be 50% lower for ET covers than for conventional covers. O&M costs for an ET cover, however, depend heavily on site-specific factors such as the need for light irrigation of vegetation, nutrient additions, erosion and biointrusion controls, and related field work. Applications often involve higher energy consumption associated with increased O&M activity. These systems are anticipated to cover many small landfills in arid or semi-arid climates over the coming decade, particularly on military properties. [55]

Profile: Fort Carson, Colorado Springs, CO

Cleanup Objectives: Contain a 15-acre hazardous waste landfill

Green Remediation Strategy: Installed a four-foot-thick monolithic ET cover

- Applied biosolids from an onsite wastewater treatment plant
- Installed a layer of straw mulch to prevent erosion
- Revegetated with native prairie grass resistant to drought and disease - Provided uncompacted soil more conducive to plant growth than conventional earthen covers

Results:

- Reduced potential for desiccation
- Reclaimed sludge otherwise destined for landfill disposal
- Enhances visual aesthetics contrasting to adjacent asphalt cover
- Saved nearly $1.5 million in construction costs compared to a conventional cover
- Incurs annual O&M costs averaging $75,000, relatively higher than conventional covers

Property End Use: Open space (McGuire, et. al., 2001)

Engineered Wetlands

Wetlands serve as biofilters capable of removing solid or dissolved-phase contaminants from ground water via passage of water through the system, while using no external sources of energy. Engineered wetlands are semi-passive networks of constructed cells specifically designed to treat contaminants in surface and/or ground water. Engineered systems accelerate cleanup through use of auxiliary components for increased control and monitoring of the treatment cells, and consequently carry higher extrinsic energy demands.

Wetlands contain rich microbial communities housed in sediment typical of marsh or swamps. In addition to biodegrading contaminants, engineered wetlands can eliminate discharge to a water treatment plant, create habitats important to healthy ecosystems, and enhance visual aesthetics of a degraded site through addition of greenspace.

Traditionally, natural or engineered wetland applications were limited to treatment of stormwater and municipal wastewater. Increased demand for wetland-based treatment systems has resulted in technology advancements enabling applications for acid mine drainage, treatment process wastewater, and agricultural waste streams. Evaluation and preliminary design of engineered wetlands as a cleanup remedy requires early assessment of site-specific characteristics and remediation/reuse goals:

- Confirming anticipated site reuse and determining whether use is compatible with a sustainable wetland,
- Estimating the time needed to establish a wetland system,
- Identifying optimal biological and chemical treatment mechanisms,
- Avoiding use of non-native, invasive, or noxious plants,
- Removing certain ground water contaminants such as mercury prior to wetland treatment, and closely monitoring the concentrations during treatment, and
- Accounting for seasonal variance in system performance and maintenance.

Designs need to account for future O&M needs, particularly for small-scale systems. If a wetland is used for buffering, rejuvenation is typically needed over time. Rejuvenation involves addition of buffering material such as limestone and removal of some sediment to maintain system grade. [56, 57]

Biowalls

A permeable reactive barrier (PRB) is an *in situ* ground water treatment technology that combines a passive chemical or biological treatment zone with subsurface fluid-flow management. PRB construction commonly involves subsurface placement of selected reactive media into one or more trenches perpendicular to and intersecting ground water flow. Passage of ground water through the barrier is driven by the natural hydraulic gradient, requiring no external energy.

Profile: British Petroleum Site, Casper, WY

Cleanup Objectives: Remediate gasoline- contaminated ground water for 50 to 100 years

Green Remediation Strategy: Installed an engineered, radial-flow constructed wetland system

- Designed wetland treatment cells for subsurface location to increase operational control, reduce offensive odors and insects, and avoid disruption of surface activity
- Constructed treatment beds of crushed concrete reclaimed from demolition of the site's former refinery
- Insulated each treatment cell with a six- inch layer of mulch to withstand temperatures reaching -35° F
- Installed native, emergent wetland plants such as bulrushes, switch grass, and cordgrass in each treatment cell
- Employed "Smart Growth" principles to complement site conversion for mixed use

Results:

- Treats up to 700,000 gallons of contaminated ground water each day
- Achieves non-detectable concentrations of benzene and other hydrocarbons - Operates year-round despite cold climate
- Incurred construction costs totaling $3.4 million, in contrast to $15.9 million for the alternative P&T system employing air stripping and catalytic oxidation

Property End Use: Office park and recreation facilities including golf and kayak courses (Wallace, 2004)

PRBs employing organic material as reactive media, otherwise known as "biowalls," are used to treat ground water containing chlorinated solvents and other organic contaminants. Reactive media typically comprise readily available, low-cost materials such as mulch, woodchips, or agricultural byproducts mixed with sand. Enhanced microbial activity within the organic material stimulates contaminant biodegradation as water slowly passes through the barrier. Sequential breakdown of contaminants results in both aerobic and anaerobic zones of the treatment area.

Biowall installation involves varying degrees of soil excavation and field mobilization, depending on site and contaminant characteristics. Typical biowall dimensions are 1 .5-3 feet in width and 25-35 feet in depth, with variable length to accommodate width of the contaminant plume. Configurations could involve a single continuous trench or a series of trenches angled for maximum plume capture. Once installed, biowalls require little field work beyond routine monitoring. Periodic replenishment of the reactive medium can be accomplished by injecting soluble organic substrate such as common soybean oil. Due to the low cost of organic materials, biowalls can be installed for one-fourth to one-third the cost of PRBs using zero valent iron, a commonly used reactive medium. [58]

Operating on the same principles as a biowall, a "bioreactor" additionally integrates a recirculation system to transfer downgradient water to the trench filled with organic media. Nutrient-rich leachate exiting the bioreactor is transferred continuously to the aquifer. Ground water pumping from the collection trench can be powered by renewable energy sources due to the low rate of water exchange required.

Monitored Natural Attenuation

Monitored natural attenuation (MNA) relies on nature's biological, chemical, or physical processes to reduce the mass, toxicity, mobility, volume, or concentration of contaminants in environmental media under favorable conditions. MNA uses an in situ approach involving close control and monitoring to achieve remediation objectives within a reasonable time frame. MNA processes include biodegradation, dispersion, dilution, sorption, volatilization, radioactive decay, and chemical or biological stabilization, transformation, or destruction of contaminants; degradation or destruction is preferred.

Profile: Altus Air Force Base, OK

Cleanup Objectives: Biodegrade a VOC hotspot 10-18 feet below ground surface in a remote location

Green Remediation Strategy: Installed a 10,000-square-foot subsurface biowall of organic material

- Filled trenches with woody waste supplied by a local municipality and cotton gin trash obtained from the local cotton industry
- Relied exclusively on power from a 200-watt PV array to recirculate ground water
- Employed a small submersible pump designed for solar applications and suitably sized for low rates of ground water transfer

Results:

- Demonstrates continued biodegradation of VOCs
- Transfers 1,300 cubic meters of carbon-enriched leach ate into the aquifer each year
- Maintains a ground water flow rate of 928 gallons each day
- Avoided significant cost for connection to the electricity grid
- Incurred capital costs of only $2,300 for the pump/solar system
- Provided a low-maintenance alternative for potentially extended cleanup duration
- Provides opportunity of re-using solar equipment (with 20- to 30-year lifespan) at other locations or sites

Property End Use: Continued military operations
(U.S. EPA/OSWER, 2007(c))

MNA is suited for sites with low potential for contaminant migration and where application ensures that all remedy selection criteria are met. MNA can be combined with aggressive remediation measures such as ground water extraction and treatment or used as a polishing step following such measures. Advantages of MNA generally include:

- Less remediation-generated waste, reduced potential for cross-media transfer of contaminants, and reduced risk of onsite worker exposure to contaminants,
- Less environmental intrusion and smaller treatment-process footprints on the environment, and
- Potentially lower remediation costs compared to aggressive treatment technologies.

When compared to aggressive treatment systems, potential disadvantages of MNA include:

- More complex and costly site characterization, longer periods needed to achieve remediation objectives, and more extensive performance monitoring (with associated energy consumption),
- Continued contamination migration or renewed contaminant mobility caused by hydrologic or geochemical changes, and

- Institutional controls to ensure long-term protectiveness and more public outreach to gain acceptance. [59]

SECTION 5. TOOLS AND INCENTIVES

Growing numbers of tools and incentives are available to site remediation and redevelopment managers for planning, financing, and implementing green projects. Several programs within EPA's ***Clean Energy*** initiative provide technical assistance and policy information, foster creation of public/private networks, and formally recognize leading organizations that adopt clean energy policies and practices. [http://www.epa.gov/cleanenergy]

- The ***Green Power Partnership*** helps organizations to buy green power designed to expand the market of environmentally preferable renewable energy sources. [http://www.epa .gov/greenpower]
- ***State Utility Commission Assistance*** is offered to utility regulators exploring increased use of renewable resources for energy production, energy efficiency, and clean-distributed generation such as co-generated heat and power. [http://www.epa.gov/cleanenergy/energyprograms/suca.html]
- The ***National Action Plan for Energy Efficiency*** engages public/private energy leaders (electric and gas utilities, state utility regulators and energy agencies, and large consumers) to document a set of business cases, BMPs, and recommendations designed to spur investment in energy efficiency. [http://www.epa.gov/cleanenergy/energy-programs/napee/index. html]
- The ***Clean Energy-Environment State Partnership Program and Clean Energy-Environment Municipal Network*** support development and deployment of emerging technologies that achieve cost savings through energy efficiency in residential and commercial buildings, municipal facilities, and transportation facilities. [http://www.epa.gov/cleanenergy/energy-programs/state-andloca l/index. html]

EPA's ***Environmentally Responsible Redevelopment and Reuse*** ("ER3") Initiative uses enforcement incentives to encourage developers, property owners, and other parties to implement sustainable practices during redevelopment and reuse of contaminated sites.

[http://www.epa. gov/compl iance/cleanup/redevelop/er3/]

As lead agency for federal energy policy, DOE continues to expand and establish new programs aimed at reducing the use of non-renewable energy sources and increasing energy efficiency.

- EERE offers grants or cooperative agreements to industry and outside agencies for renewable energy and energy efficiency research and development. Assistance is available in the form of funding, property, or services. In fiscal year 2004, EERE awarded $506 million in financial assistance. [http://www1.eere.energy.gov/financing/types_assistance. html]

- EERE also provides grants to state energy offices for energy efficiency and renewable energy demonstration projects as well as analyses, evaluation, and information dissemination. [http ://www.eere .energy. gov/state_energy_program/]

State and local mechanisms are evolving quickly to meet national energy goals for the coming decades. State renewable energy portfolios help meet these goals by offering (1) third-party funding mechanisms that support public/private partnerships for generation of electricity from renewable resources, (2) reduced purchasing rates for electricity generated from renewable resources, and (3) tax credits for energy production from renewable resources. The ***Database of State Incentives for Renewables and Efficiency*** (DSIRE) provides quick access to information about renewable energy incentives and regulatory policies administered by federal and state agencies, utilities, and local organizations. Information is updated frequently through a partnership among the North Carolina Solar Center, the Interstate Renewable Energy Council, and DOE. [www.dsireusa.org/]

State authorities are working with commissioned utilities to develop a host of tools and incentives for using green practices. Programs in Minnesota and California demonstrate some of the mechanisms becoming available.

- The Minnesota Pollution Control Agency (MPCA) ***Green Practices for Business, Site Development, and Site Cleanups: A Toolkit*** provides online tools to help organizations and individuals make informed decisions regarding sustainable BMPs for use, development, and cleanup of sites. [http://www. pca .state. mn . us/programs/p2 -s/toolkit/index. html]
- The ***State of California Self-Generation Incentive Program*** *(SGIP)* provides incentives for installation of renewable energy systems and rebates for systems sized up to 5 MW. Qualifying technologies include PV systems, microturbines, fuel cells, and wind turbines. [http://www.pge.com/selfgen/]

SECTION 6. FUTURE OPPORTUNITIES

Significant opportunities exist to increase sustainability of site remediation while helping to meet national, regional, and state or local goals regarding natural resource conservation and climate change. Decision-makers are encouraged to take advantage of newly demonstrated or emerging technologies and techniques in ways that creatively meet the objectives of site cleanup as well as revitalization. Effective green remediation can provide a range of new opportunities.

- **Building Stronger Communities**
    - Renew or form new partnerships among organizations and individuals with common environmental, economic, and social concerns, including energy independence,
    - Identify optimal methods that stakeholders can use to influence the direction of remediation and revitalization and to maintain an active voice throughout a project, and

- Work more efficiently with local engineering firms involved in cleanup design, construction, and operations.
- **Expanding the Options for Site Reuse**
    - Evaluate options presented by a larger universe of potential developers,
    - Identify new solutions for unresolved site issues, and
    - Facilitate new incentives for current site owners.
- **Increasing Economic Gains**
    - Integrate new energy-related businesses into local and regional infrastructures,
    - Demonstrate specific technical needs to be met by commercial product and service vendors, and
    - Foster government initiatives that reward businesses employing sustainable practices.
- **Increasing Environmental Benefits of Cleanups**
    - Enhance environmental conditions beyond immediate target areas,
    - Participate in state and local initiatives collectively working to meet goals for natural resource and energy conservation, and
    - Showcase more sustainable cleanup and revitalization strategies that readily apply to other sites.

Additional information on opportunities and tools for implementing green remediation is frequently uploaded to the EPA Office of Superfund Remediation and Technology Innovation's CLU-IN Web page on ***Green Remediation*** (http://www.cluin.org/green remediation). Future electronic updates to this primer also will be available on CLU-IN to share emerging information on green remediation.

SECTION 7: REFERENCES

ACORE (American Council on Renewable Energy) and ABA (American Bar Association).(2008). January 16, *Renewable Energy Seminar and Teleconference.*

Capital Research. (2005). Legislative History *Report: H.R. 6 - the Federal Energy Policy Act of 2005.* http://www.capitolresearch.us/ reports/2005energy_policy.html#exhibits.

Executive Order 13423. (2007). January 24, *Strengthening Federal Environmental, Energy, and Transportation Management.* http://www.ofee.gov/.

McGuire, Patrick, E., John, A., England, Brian, & Andraski, J. (2001). "An Evapotranspiration Cover for Containment at a Semiarid Landfill Site," *Proceedings of International Containment and Remediation Technology Conference and Exhibition.*

Solid Waste Association of North America. (2005). Jeremy O'Brien, Applied Research Foundation. *Comparison of Air Emissions from Waste-to-Energy Facilities to Fossil Fuel Power Plants.*

U.S. DOE/EERE. July (2007). Federal Energy Management Program. Fact Sheet: *Super ESPC – Just the Facts: Energy Savings Performance Contracting.* http://www1. eere.energy.gov/femp/program/equip_ procurement.

U.S. DOE/Energy Information Administration (EIA). July (2007). *Annual Energy Review 2006, Diagram 6: Electricity Flow, 2006.* http://www.eia.doe.gov/aer.

U.S. DOE/EIA. April (2008). *Municipal Solid Waste: Table 7, Waste Energy Consumption by Type and Energy Use Sector, 2005*. http://www.eia.doe.gov/cneaf/solar.renewables/page/trends/table7.html.

U.S. DOE/NREL. March (2008). (a). *Solar Energy Resource Atlas of the U.S.: Annual Average Wind Resource Estimates*. Compiled by Billy Roberts, NREL.

U.S. DOE/NREL. March (2008). (b). *Wind Energy Resource Atlas of the U.S.: 50 Meter Wind Power Resource*. Compiled by Billy Roberts, NREL.

U.S. EPA. (2008). [online]. *Clean Energy: Municipal Solid Waste*. http://www.epa.gov/cleanenergy/energy-and-you/affect/municipal-sw. html.

U.S. EPA/National Center for Environmental Innovation. March (2006). *Sectors Program: Performance Report 2006*. EPA 100-R-06-002. http://www.epa.gov/ispd/performance.

U.S. EPA/Office of the Chief Financial Officer. September (2006). *2006-2011 EPA Strategic Plan*. http://www.epa.gov/ocfo/plan/plan.htm.

U.S. EPA/OSRTI. November (2005). Cost and Performance Report: *Phytoremediation at Naval Air Station – Joint Reserve Base Fort Worth, Fort Worth, TX.*

U.S. EPA/OSWER. October (1997). *Innovative Uses of Compost: Composting of Soils Contaminated by Explosives*. EPA 530-F-97-045. http://www.epa.gov/epaoswer/non-hw/compost.

U.S. EPA/OSWER. December (2002). *Elements for Effective Management of Operating Pump and Treat Systems*. EPA 542-R-02-009. http://clu-in.org/techfocus/default.focus/sec/Remediation_ Optimization/ cat/Guidance/page/3/.

U.S. EPA/OSWER. (2004). *Cleaning Up the Nation's Waste Sites: Markets and Technology Trends*. EPA 542-R-04-015. http://www.cluin.org/market/.

U.S. EPA/OSWER. December (2006). Final Report, Pilot Region-Based Optimization Program for Fund-Lead Sites in EPA Region 3: *Site Optimization Tracker, Havertown PCP Site, Havertown, Pennsylvania*. EPA 542-R-06-006j. http://clu-in.org/search/default.cfm?search term = Havertown + PCP&t=al l&advl it= 0.

U.S. EPA/OSWER. August (2007a). *Green Remediation and the Use of Renewable Energy Sources for Remediation Projects*. Amanda Dellens, National Network for Environmental Management Studies Fellow, Case Western Reserve University. http://cluin.org/s.focus/c/pub/i/1474/.

U.S. EPA/OSWER. May (2007b). "Wind Turbine Cost Study Shows Need for Redesigned Ground- Water Remediation Systems," *Technology News and Trends*. EPA 542-N-06-009. http://clu-in.org/products/newsltrs/tnandt/view.cfm?issue=0507.cfm#1.

U.S. EPA/OSWER. May (2007c). "Solar Power Recirculates Contaminated Ground Water in Low- Energy Bioreactor," *Technology News and Trends*. EPA 542-N-06-009. http://clu-in.org/products/newsltrs/tnandt/view.cfm?issue=0507.cfm#1.

U.S. EPA/OSWER. April (2008a). "Integrated Technology Approach Used to Remediate Site Contaminated by 56 Chemicals," *Technology News and Trends*. EPA 542-N-08-002. http://cluin.org/products/ newsltrs/tnandt/.

U.S. EPA/OSWER. April (2008b). *Energy Consumption and Carbon Dioxide Emissions at Superfund Cleanups. http://www.cluin. org/greenremediation.*

U.S. EPA/OSWER. April (2008c). Green Remediation online. *Profiles of Green Remediation: Former St. Croix Alumina, St. Croix, VI.* http://clu-in.org/green remediation.

U.S. EPA/Office of Water. December (2007). *Reducing Stormwater Costs through Low*

Impact Development (LID) Strategies and Practices. EPA 841-F-07-006. http://www.epa.gov/owow/nps/lid/costs07/.

University of Missouri-Rolla. (2005). *Ground Water Remediation Powered by a Renewable Energy Source*. In collaboration with EPA and U.S. Army Corps of Engineers Kansas City District. http://clu-in.org/greenremediation/tab_c.cfm.

Wallace, Scott. September/October (2004). "Engineered Wetlands Lead the Way," Land and Water. http://www.landandwater.com/features/ vol48no5/vol48no5_1.php.

SECTION 8. GENERAL RESOURCES

Expanding numbers of technical, planning, and financial resources for implementing green remediation are available from federal or state agencies, academic organizations, and sector-specific trade associations. The following documents and online resources provided key information for this primer and are readily available to readers interested in learning more about specific topics.

1. U.S. EPA online. *Sustainability*. http://www.epa.gov/sustainability
2. U.S. EPA online. *Climate Change*. http://www.epa.gov/ climatechange/
3. U.S. DOE/EERE Federal Energy Management Program. January 2008. *2007 Federal Energy Management Program (FEMP) Renewable Energy Requirement Guidance for EPACT 2005 and Executive Order 13423 Final*. http://www1.eere.energy.gov/femp/
4. U.S. DOE/EERE Federal Energy Management Program. January 2008. DOE Supplemental Guidance to the Instructions for Implementing Executive Order 13423, "Strengthening Federal Environmental, Energy, and Transportation Management." *Establishing Baseline and Meeting Water Conservation Goals of Executive Order 13423*. January 2008. http://www1.eere.energy. gov/femp/.
5. National Ground Water Association online. *Ground Water Protection and Management Critical to the Global Climate Change Discussion*. http://www.ngwa.org/PROGRAMS/government/issues/climate
6. Comprehensive Environmental Response, Compensation and Liability Act: 42 U.S.C. § 9601– 9675.
7. National Oil and Hazardous Substances Pollution Contingency Plan: 40 CFR 300(e)(9).
8. Energy Policy Act of 2005. August 8, 2005. Public Law 109-48. http://thomas.loc.gov/
9. Energy Independence and Security Act of 2007. December 19, 2007. Public Law 110-140. http://thomas.loc.gov/
10. U.S. DOE online. *Energy Efficiency and Renewable Energy*. http://www.eere.energy.gov
11. U.S. Green Building Council online. *LEED*. http://www.usgbc.org/DisplayPage.aspx?CategoryID=19.
12. U.S. DOE/EPA online. *Energy Star*. http://www.energystar.gov/
13. U.S. EPA online. *GreenScapes*. http://www.epa.gov/greenscapes/

14. Smart Growth Network online. *Principles of Smart Growth.* http://www.smart growth.org/about/principles/default.asp?res=1680
15. National Institute of Building Sciences online. *Federal Green Construction Guide for Specifiers.* http://www.wbdg.org/design/ greenspec.php.
16. General Services Administration online. *Go Green: GSA Environmental Initiatives.* http://www.gsa.gov/Portal/gsa/ep/ home.do?tabId=10.
17. U.S. EPA online. *Sector Strategies Program.* http://www.epa.gov/ ispd/
18. Piedmont Biofuels online. http://biofuels.coop/coop/
19. U.S. EPA/OSWER. May 1991. *Management of Investigation Derived Waste During Site Inspections.* OERR Directive 9345.3-02. EPA 540/G91/009. http://nepis.epa. gov/EPA/html/pubs/pubtitleoswer.htm
20. U.S. EPA/OSWER. 1992. *Guide to Management of Investigation-Derived Wastes.* Directive 9345.3-03FS. Triad Resource Center [online]. http://triadcentral.org
21. ITRC online. *Diffusion/Passive Sampler Documents.* http://www.itrcweb.org/ gd_DS.asp
21. Triad Resource Center online. http://triadcentral.org
23. U.S. EPA online. Clean Construction USA: *Construction Air Quality Language.* http://www.epa.gov/diesel/construction/contract-lang.htm
24. U.S. EPA/Region 9 online. *Cleanup – Clean Air Initiative.* http://www.epa.gov/ region09/cleanup-clean-air/
25. U.S. EPA/National Center for Environmental Innovation. March 2007. *Cleaner Diesels: Low Cost Ways to Reduce Emissions from Construction Equipment.* http://www.epa.gov/sectors/construction/
26. U.S. EPA online. Clean Energy: *Greenhouse Gas Equivalencies Calculator.* http://www.epa.gov/cleanenergy/energy-resources
27. U.S. EPA. Polluted Runoff (Nonpoint Source Pollution): *Reducing Stormwater Costs through Low Impact Development (LID) Strategies and Practices.* EPA 841-F-07-006. http://www.epa.gov/owow/nps/lid/costs07/
28. U.S. EPA online. *Effluent Limitation Guidelines.* http://www.epa.gov/water
29. U.S. EPA/OSWER. August 2007. *Integrating Water and Waste Programs to Restore Watersheds: A Guide for Federal and State Project Managers.* EPA 540K07001. http://www.epa.gov/superfund/resources
30. U.S. EPA online. *Eco Tools: Tools for Ecological Land Reuse.* http://cluin.org/ products/ecorestoration/
31. U.S. EPA online. Municipal Waste: *Reduce, Reuse, and Recycle.* http://www. epa.gov/msw/reduce.htm
32. U.S. EPA online. Clean Energy: *Power Profiler.* http://www.epa.gov/cleanenergy/ energy-andyou/how-clean. html
33. U.S. DOE/EERE online. Industrial Technologies Program: *Pumping System Assessment Tool.* http://www.eere.energy.gov
34. U.S. EPA Technology Innovation Office and U.S. Army Corps of Engineers. *CLU-IN Online Seminar: Remediation System Evaluations and Optimization of Pump and Treat Projects.* http://clu-in.org/s.focus/c/pub/i/826/
35. U.S. EPA online. *Remediation System Optimization.* http://clu-in.org/rse
36. U.S. Air Force Center for Engineering and the Environment online. *Remedial Process Optimization.* http://www.afcee.brooks.af.mil/products/rpo/default.asp

37. Federal Remediation Technologies Roundtable online. *Remediation Optimization.* http://www.frtr.gov/optimization
38. ITRC online. *Remediation Process Optimization Documents.* http://www.itrcweb.org/gd_RPO.asp
39. U.S. EPA online. Green Power Partnership: *Green Power Equivalency Calculator.* http://www.epa.gov/grnpower/pubs/calculator.htm
40. U.S. DOE/NREL online. *Science and Technology.* http://www.nrel.gov/
41. U.S. DOE/NREL online. *Solar Research.* http://www.nrel.gov/solar/
42. American Solar Energy Society online. http://www.ases.org/index.htm
43. U.S. DOE/NREL online. *Wind Research.* http://www.nrel.gov/wind/
44. American Wind Energy Association online. http://www.awea.org/faq/
45. U.S. EPA online. *Combined Heat and Power Partnership.* http://www.epa.gov/chp/
46. U.S. EPA online. *Landfill Methane Outreach Program.* http://www.epa.gov/landfill/
47. U.S. DOE/EERE online. State Activities and Partnerships: *Waste-to-Energy Projects Gain Momentum in the United States* http://www.eere.energy.gov/states/state_news_detail.cfm/news_id=10404/state=AL
48. Idaho National Laboratory online. *Geothermal Energy.* http://geothermal.inel.gov/
49. U.S. DOD Environmental Security Technology Certification Program. March 2006. *Design Document for Passive Bioventing.* ESTCP Project: ER-9715. http://www.cluin.org/techfocus/default.focus/sec/Bioventing%5Fand%5FBiosparging/cat/Guidance/
50. Savannah River National Laboratory online. *Tech Transfer: Environmental Remediation.* http://www.srs.gov/general/busiops/tech-transfer/
51. U.S. EPA online. *Carbon Sequestration in Agriculture and Forestry.* http://www.epa.gov/sequestration/
52. U.S. EPA online. *CLU-IN Technology Focus: Bioremediation of Chlorinated Solvents.* http://clu-in.org/techfocus/
53. U.S. EPA online. *CLU-IN Technology Focus: Phytoremediation.* http://clu-in.org/techfocus/default.focus/sec/Phytoremediation/ cat/Overview/
54. U.S. EPA/OSWER. December 2007. *The Use of Soil Amendments for Remediation, Revitalization, and Reuse.* EPA 542-R-07-013. http://www.clu-in.org/s.focus/c/pub/i/1515/
55. U.S.EPA/OSWER. September 2003. *Evapotranspiration Landfill Cover Systems Fact Sheet.* EPA 542-F-03-015. http://www.clu-in.org/download/remed/epa542f03015.pdf
56. U.S. EPA online. *Constructed Wetlands.* http://www.epa.gov/owow/wetlands
57. ITRC online. *Constructed Treatment Wetlands Documents.* http://www.itrcweb.org/gd_CW.asp
58. U.S. Naval Facilities Engineering Service Center, Environmental Restoration Technology Transfer (ERT2) online. *Permeable Mulch Biowalls.* 58. http://www.ert2.org/PermeableMulchBiowalls/ tool.aspx
59. U.S. EPA online. *CLU-IN Technology Focus: Natural Attenuation.* http://clu-in.org/techfocus/default.focus/sec/Natural%5FAttenuation/ cat/Overview/

CHAPTER SOURCES

The following chapter have been previously published:

Chapter 1 – This is an edited, reformatted and augmented version of a United States Environmental Protection Agency publication, report EPA 542-R-08-003, dated February 2009.

Chapter 2 – This is an edited, reformatted and augmented version of a United States Environmental Protection Agency publication, report EPA 542-R-08-002, dated April 2008.

INDEX

A

B

C

D

E

F

G

H

N

O

P

Q

R

S

T

U

V

W

Z